Estates Gazette

Planning Law Reports

1991
Volume 2

Edited by
Barry Denyer-Green LLM PhD FRICS
of the Middle Temple, barrister
and
David Lamming LLB LLM
of Gray's Inn and Lincoln's Inn, barrister

Audrey Boyle BA
Publisher's Editor

1991
Estates Gazette
A member of Reed Business Publishing Group

Routledge
Taylor & Francis Group
New York London

EG Books is an imprint of Routledge

2 Park Square, Milton Park, Abingdon, Oxon OX14 4RN
711 Third Avenue, New York, NY 10017, USA
Routledge is an imprint of Taylor & Francis Group, an Informa business.

First published 1991
ISBN for complete set of 1991 volumes: 0 7282 0165 8
ISBN for this volume: 0 7282 0167 4

ISSN 0954-1977

Transferred to Digital Printing 2007

Typesetting by Digital Graphics, 147-149 Wardour Street, London

PLANNING LAW REPORTS

[1991] 2 PLR

CASES REPORTED IN VOLUME 2

NOTER-UP

INDEX OF SUBJECT-MATTER

CASES JUDICIALLY CONSIDERED

For case references see pages shown in bold type **PAGES**

Harvey *v* Secretary of State for Wales and another

COURT OF APPEAL

Mustill and Farquharson LJJ and Sir Michael Kerr

November 8 1989

Planning permission for erection of bungalow — Condition requiring existing bungalow to be demolished or occupied by agricultural worker — Condition not complied with — Enforcement notice issued more than four years after date for compliance — Whether condition "relates to the carrying out of operational development" — Whether enforcement notice valid A

Following the grant of planning permission in November 1961 a bungalow ("the new bungalow") was erected on Goitre Farm, St Mellons, B Monmouthshire. In September 1974 planning permission was granted for a second bungalow ("the replacement bungalow") on the farm subject to a condition that the new bungalow be demolished or used for the purpose of agriculture other than as a dwelling within one month of the beneficial use of the replacement bungalow. The replacement bungalow was constructed and occupied by November 5 1977. The new bungalow was sold and eventually came to be occupied by the appellant; the condition was never complied with. C

In September 1985 the second respondents, Cardiff City Council, issued an enforcement notice alleging a breach of planning control by reason of the failure to comply with the planning condition. The appellant appealed contending that the enforcement notice was invalid by reason of having been issued more than four years after the alleged breach: see section 87(4)(*b*) of the Town and Country Planning Act 1971 (section 172(4)(b) of the Town and Country Planning Act 1990). The appellant's appeal against the decision of the inspector to uphold the enforcement notice was dismissed by Sir Frank D Layfield QC (sitting as a deputy judge of the Queen's Bench Division). The appellant appealed contending that the condition related to operational development and the enforcement notice had therefore been served after the time-limit of four years.

Held The appeal was allowed.

For the purposes of section 87(4)(*b*) of the Town and Country Planning Act E 1971, the real question is whether any condition or limitation relates to the carrying out of operational development and not whether the condition required some operational development: see p 5A. In the present case planning permission was granted for the erection of the replacement bungalow; this was a building operation, section 87(4)(*b*) was satisfied and the enforcement notice was invalid by being served after the expiration of the four year time-limit. For the purposes of this subsection the condition or limitation does not itself have to contain or involve operational development: see p 7. F

Cases referred to in the judgments

Newbury District Council v *Secretary of State for the Environment* [1981] AC 578; [1980] 2 WLR 379; [1980] 1 All ER 731; (1980) 78 LGR 306; 40 P&CR 148; [1980] JPL 325, HL

Peacock Homes Ltd v *Secretary of State for the Environment* (1984) 83 LGR 686; 48 P&CR 20, CA G

Pyx Granite Co Ltd v *Ministry of Housing and Local Government* [1958] 1 QB 554; [1958] 2 WLR 371; [1958] 1 All ER 625; (1958) 56 LGR 171; 9 P&CR 204, CA

Appeal against decision of Sir Frank Layfield QC

This was an appeal from a decision of Sir Frank Layfield QC (sitting as a deputy judge of the Queen's Bench Division), who had dismissed an appeal against the

A decision of the first respondent, the Secretary of State for Wales, by his inspector, who had dismissed an appeal against an enforcement notice issued by the second respondents, Cardiff City Council.

Wyn Williams (instructed by Hopkins & Wood, for Edwards Geldard, of Cardiff) appeared for the appellant, Frederick Harvey.
B *Richard Drabble (instructed by the Treasury Solicitor) appeared for the first respondent, the Secretary of State for Wales.*
The second respondents, Cardiff City Council, did not appear and were not represented.

The following judgments were delivered.

MUSTILL LJ: I will ask Sir Michael Kerr to give the first judgment.

C **SIR MICHAEL KERR:** This is an appeal from Sir Frank Layfield QC (sitting as a deputy judge of the Queen's Bench Division), arising from a judgment delivered on July 26 1988. It relates to certain buildings on land known as Goitre Farm, St Mellons, [formerly] in Monmouthshire. The court has been provided with a statement of facts which are not in issue and it is convenient to summarise the history.

There was a farmhouse on this land prior to 1961, and in November 1961 the
D owner, Mr T B Evans, obtained planning permission for the erection of a bungalow with a garage for an agricultural worker. Pursuant to this permission a bungalow and garage were constructed and were thereafter occupied by Mr L H Evans, the son of Mr T B Evans. The building was subsequently referred to as the "new bungalow".

That remained the position until the early 1970s when a new highway known
E as A48(M) was under construction. It became clear to the highways department of the Welsh Office that this would adversely affect the new bungalow because of its proximity to it. Indeed, when the highway had been constructed its edge was within 10m or so of the new bungalow boundary.

In May 1974 Mr T B Evans (Mr Evans snr) thereupon sought two further planning permissions. First, he wanted to build a replacement farmhouse. Second, he wanted to build a bungalow (subsequently called the "replacement
F bungalow") in place of the new bungalow. Both permissions were granted by the Cardiff City Council on September 11 and 12 1974. But for the purpose of this case we are concerned with the permission to build the replacement bungalow on the latter date. It recites Mr Evans' application for planning permission to build a dwellinghouse "in place of dwelling adversely affected by proposed motorway", and then goes on:

> G THE CARDIFF CITY COUNCIL, . . ., in pursuance of its power under the Town and Country Planning Act 1971 hereby PERMITS the development to be carried out . . . subject to compliance with the conditions specified hereunder.

One condition, (e), to which Mr Drabble has drawn attention, provided that the replacement bungalow was to be occupied only by a person employed in agriculture. However, the condition with which we are concerned is condition
H (f):

> (f) The existing farmhouse shall be demolished or used for the purpose of agriculture other than a dwelling to the satisfaction of the Local Planning Authority, and within one month of the beneficial use of the dwelling hereby approved.

What that means is that the new bungalow was to be demolished or used for agricultural purposes other than as a dwelling and that that was to happen within one month from the occupation of the replacement bungalow whose erection was approved by this permission.

The permitted development was fully carried out. A new farmhouse and the replacement bungalow were built. Mr L H Evans (Mr Evans jun) occupied one and his, by then widowed, mother occupied the other. The important date for present purposes is that Mr Evans took up residence in the replacement bungalow on November 5 1977. It therefore follows from the wording of condition (f) that by December 5 1977 the original new bungalow had to be either demolished or used for agricultural purposes. In the event neither occurred. It is accordingly common ground that by about December 5 1977 there was a breach of condition (f) attached to this planning permission. That, of course, leads to the question whether the subsequent enforcement notice was served in time. But before coming to that I will complete the history.

The new bungalow remained empty and in July 1980 it was purchased by the Welsh Office. By that time views had begun to fluctuate as to the purposes for which it might properly be used, and there was no longer any suggestion that it should be demolished. It remained unoccupied and unused until it was sold by auction by the Welsh Office in October 1982. It was then bought by the present appellant, Mr Harvey, who took up residence in it in early 1984. That led to the serving of a number of enforcement notices in 1985, because Mr Harvey was using the bungalow for residential purposes contrary to one of the conditions attached to the planning permission which Mr Evans had obtained.

Three enforcement notices called A, B, and C were served in somewhat different terms. But we are concerned only with notice B, which was upheld. It was issued, as were the others, on September 4 1985 — long after four years from the date of the breach. The operative paragraph of the notice was in the following terms:

> . . . on the 12th September 1974 the Council as Local Planning Authority . . . granted permission for the development of land at Goitre Farm, St Mellons, Cardiff by the carrying out of the building operations described in Schedule 2(1) below. The breach of planning control which appears to have taken place consists in the failure to comply with the condition or limitation subject to which the said permission was granted, described in Schedule 2(2).

Schedule 2(1), headed "Details of planning permission", says:

> Permission to build dwellinghouse in place of dwelling (New Bungalow) adversely affected by . . . motorway.

Schedule 2(2), headed "Conditions which have not been complied with", recited condition (f), which I have already set out. Schedule 3, setting out the steps required to be taken pursuant to the enforcement notices, required that the use of the premises for residential purposes should cease. Reasons were given why the notice was served. These concerned the undesirability of sporadic residential development in an agricultural area, noise levels from the adjacent motorway, and other matters.

The three enforcement notices were challenged and a public inquiry was held in May 1986. The report of the inspector quashed notices A and C but upheld notice B. The matter then went to the Secretary of State for Wales and by a decision letter dated June 25 1987 he upheld notice B. The applicant then issued a notice of motion challenging that decision. This came before Sir Frank Layfield QC, who dismissed the notice of motion. Mr Harvey now appeals from that decision.

It is right to stress that a substantial number of issues were argued below which have not survived. The only ones which have survived for the purpose of this appeal are, first, the question of the time-limit of the enforcement notice, which was, of course, also argued below, and, second, a matter on which we

A found it unnecessary to hear Mr Williams, whether the decision of the Secretary of State to uphold the enforcement notice B could be criticised on *Wednesbury* grounds. We have not heard him on this because we formed the view, after hearing both counsel, that the enforcement notice was served out of time. The argument on this issue was the same before us as it had been below; this was common ground. But its real impact is not reflected in the judgment, perhaps because there were a number of other issues.

B In order to see the point one has to go to section 87 of the Town and Country Planning Act 1971. It is common ground that if the four-year time-limit prescribed by that section is applicable, then enforcement notice B was out of time, since it was served well after four years from the date of the breach, which it is accepted occurred in December 1977, as I have already mentioned. Section 87 defines an enforcement notice and subsection (3) is in the following
C terms:

> (3) There is a breach of planning control —
> (*a*) if development has been carried out, whether before or after the commencement of this Act, without the grant of the planning permission required in that behalf . . . ; or
> (*b*) if any conditions or limitations subject to which planning permission was granted have not been complied with.

D We are concerned with (3)(*b*) for the reasons I have already explained. Subsection (4) provides as follows so far as material:

> (4) an enforcement notice which relates to a breach of planning control consisting in —
> (*a*) the carrying out without planning permission of building, engineering, mining or other operations in, on, over or under land; or
> (*b*) the failure to comply with any condition or limitation which relates to
> E the carrying out of such operations and subject to which planning permission was granted for the development of that land;
> . . .
> may be issued only within the period of four years from the date of the breach.

What has been argued on behalf of Mr Harvey, but is not reflected in the judgment, is that subsection (4)(*b*) is satisfied in the following way. The crucial question which one has to ask oneself, since there was admittedly a failure to
F comply with a condition, is whether the condition — in this case the demolition of the new bungalow, or alternatively its use for agricultural purposes only, within a prescribed time — is one "which relates to the carrying out of such operations and subject to which planning permission was granted for the development of that land". Does the condition, requiring either demolition or agricultural use, relate to the carrying out of operations which can, in short, be referred to as "operational development"? That is the key
G question. The learned judge, in the way in which he understood the argument addressed to him, answered that question in the negative, because he considered only the contents of the condition. These were either demolition or agricultural use, neither of which he regarded as operational development. To that extent his judgment is perfectly correct. He said at p 10, letter E of the transcript:

> H The circumstances of this permission do not I consider fall within "operational development". The condition itself requires one of two courses of action to be taken. It first requires "demolition", which I take not to be operational development. Second, it requires that a particular use be instituted or continued. The latter requirement I consider does not fall within the terms of the subsection.

Accordingly, what the learned judge was considering, and considering only, was the question of whether the condition required some operational

development. But that is not the real point. The real question is whether any A
condition or limitation "relates to the carrying out of" what can shortly be
described as "operational development". Here it is said that the condition
that — within the time prescribed — the new bungalow be either demolished or
used only for agricultural purposes clearly related to operational development
for which planning permission was granted, namely the construction of the
replacement bungalow.

Put in that way, it seems to me that there is simply no answer, both as a B
matter of first impression of the wording of the section on the facts of this case
and, indeed, in the light of the terms of the planning permission and of
condition (f) itself.

What appears to have happened is that the arguments and the judge's
consideration of the case were overshadowed by the only reported decision on
this provision — the decision of this court in *Peacock Homes Ltd* v *Secretary* C
of State for the Environment (1984) 83 LGR 686. The case is concerned with
the lengthy history of a warehouse building in Epsom. It was originally
permitted to be erected as long ago as July 1950 but subject to a condition that,
unless the time were extended, it would be demolished by the end of 1950.
However, that permission was extended on about six occasions, and the final
extension did not expire until March 31 1974. An enforcement notice was then
served more than four years later on the ground that the building had not been D
demolished by the termination of the final extension. The question was
whether section 87(4)(*b*) applied to those facts, in which event the enforcement
notice was out of time. This court held that the enforcement notice was out of
time because, to paraphrase the ratio of the decision, the original erection of
the bungalow had been a permission for operational development, and the
condition of demolition by a certain time — albeit not itself operational
development — was one which related to a permission for operational E
development, the original erection of the building. It was also held that for the
purpose of arriving at that conclusion it was permissible to treat all the
extensions over a period of some 24 years as a single period.

So what one had there was planning permission for operational
development, the erection of the warehouse building, subject to a condition
which was not a condition requiring or involving any operational F
development, namely the demolition of the building, it being expressly held
that demolition does not constitute operational development. But the court
nevertheless held that the relationship between the non-operational
condition, ie demolition, and the original permission for operational
development in the form of the erection of the building satisfied section
87(4)(*b*), with the result that the enforcement notice was out of time.

The leading judgment was given by Dillon LJ, who reviewed the history of G
section 87 by pointing out that, whereas before 1968 any enforcement notice
had to be served within four years, the time-limit thereafter applied only to
situations in which it would or should be relatively easy for local authorities to
discover whether there had been a breach of planning control or a breach of a
condition relating to planning permission. He dealt with that at p 690. At p 691
he said in relation to the wording of para (*b*):

> The earlier words in paragraph (*b*), "any condition or limitation", are H
> limited in their scope in that they apply only to conditions or limitations
> which relate "to the carrying out of such operations", namely, building,
> engineering, mining or other operations in, on, over or under land, or in other
> words "operational development". They do not extend to all conditions
> which might be imposed on a grant of any planning permission.

What he had in mind there I think, having regard to the argument so far as I remember it from the case, was that planning permissions merely involving changes of use, as opposed to operational development, would not fall within the time-limit of section 87. He then discussed the submissions advanced on behalf of the Secretary of State to the effect that the time-limit did not apply to the somewhat special facts of that case. He rejected those submissions at p 693 in three paragraphs which contain the ratio of the decision. He said:

> These anomalous distinctions lead me to wonder whether the Secretary of State's interpretation of section 87(4)(*b*) can really be correct or whether there is scope for a wider interpretation which would avoid such anomalies. The key words "which relates to" in the phrase "a condition or limitation which relates to the carrying out of such operations", are not words of art. They are words often used in the Act, as in other statutes, in a very general sense, eg, in the opening words of section 32(1) and 87(4) of the Act. They may often mean little more than "which has to do with".
>
> Whatever else the phrase may cover, it is appropriate, in my judgment, to cover the situation in which a planning permission for a limited period is granted for the erection of a building and its retention for the limited period subject to a condition that at or before the end of the period the building will be removed. Such a condition can fairly be said, in my judgment, to relate to the erection of the building, which was only authorised for the limited period.
>
> So, equally, if the limited period of the original permission for the erection of the building is extended for a further period or periods subject to the same or a similar condition, the condition can fairly be said, in my judgment, to relate to the erection of the building. It matters not, as it seems to me, whether the time limit of the condition is the original time limit or the time limit as extended by one or more subsequent permissions.

In the present case, we are, of course, not concerned with any similar history. It is also to be noticed that in that case the permission for operational development, namely the permission to erect the building, related to the building which was itself the subject-matter of the enforcement notice. In the present case, what is said to be the permitted operational development does not relate to the subject-matter of the enforcement notice, the new bungalow, but to the erection of a different building, the replacement bungalow. However, I cannot see that that makes any difference. Indeed, as I follow Mr Drabble's argument, that was not his contention. But before I come to his contention I should just mention that Mr Williams also helpfully drew attention to the decision of the House of Lords in *Newbury District Council* v *Secretary of State for the Environment* [1981] AC 578 in which it was restated that a condition attached to a grant of planning permission is valid only if it relates to a permitted development, and Mr Williams submits that "relates to" in the present case has the same wide meaning. He drew attention to a passage in the speech of Viscount Dilhorne at p 599 where he cites the well-known judgment of Lord Denning in *Pyx Granite Co Ltd* v *Ministry of Housing and Local Government* [1958] 1 QB 554 where he said at p 572:

> Although the planning authorities are given very wide powers to impose "such conditions as they think fit," nevertheless the law says that those conditions, to be valid, must fairly and reasonably relate to the permitted development.

That was the phrase which Viscount Dilhorne approved.

However, what Mr Drabble argued, if I may say so very lucidly, is that, despite the remarks of Dillon LJ in the *Peacock* case which made it clear that the words "which relates to" have a wide meaning and may mean little more than "has to do with", on its true construction section 87(4)(*b*) has a much narrower scope. He emphasised the words "carrying out" in the phrase "which relates to the carrying out of such operations". He has to accept that in the present case the condition of demolition or agricultural use of the new bungalow relates to the operational development permitted by the construction of the replacement bungalow. But what he says is that it does not relate to the "carrying out" of that operational development — that is to say, the actual construction of the building. I do not myself accept that that invalidates the argument which has been presented to us, and there is certainly nothing in the decision in the *Peacock* case which supports Mr Drabble's narrow construction. He recognises that himself, but submits that the first passage which I quoted from the judgment of Dillon LJ at p 691 at any rate leaves it open that there may be some distinction to be drawn on the lines which he suggests. However, I cannot attach any such importance, as he would have, to the words "carrying out" in the phrase "which relates to the carrying out of such operations". As Mr Williams pointed out in his reply, when one looks, for instance, at the definition of "development" in section 22 of the Act one can see that "development" is generally used in the context of the words "carrying out". Section 22(1) provides:

> In this Act, except where the context otherwise requires, "development", . . . means the carrying out of building, engineering, mining or other operations

ie the carrying out of what are compendiously known as "development operations" or "operational development".

I have already read the planning consent which expressly grants permission to carry out the operation of building the replacement bungalow subject to condition (f), which I have also already set out. Therefore it seems to me that subsection 4(*b*) is satisfied precisely. If anything, what Dillon LJ said at p 693 in the *Peacock* case, about the width of the expression "which relates to", supports Mr Williams' argument.

The learned judge, as I have explained, for some reason — probably because he had many other issues to deal with — did not focus on that way of presenting the case, and it is not discussed in his judgment. He appears to have thought that section 87(4)(*b*) requires that the broken condition or limitation must itself contain or involve some operational development. But this is not so. The condition or limitation merely has to relate to some (other) operational development for which permission was granted subject to compliance with the condition or limitation, whatever it may be.

Accordingly, I would allow this appeal and hold that the enforcement notice was served out of time and that it is therefore invalid.

FARQUHARSON LJ: I agree.

MUSTILL LJ: I also agree.

Appeal allowed with costs.

East Hampshire District Council *v* Davies

COURT OF APPEAL
Fox and Butler-Sloss LJJ
November 20 1990

A **Injunction — Use of land in breach of planning control — Enforcement notices served but under appeal — Whether appropriate case for grant of interlocutory injunction**

The defendant, William Davies, was the owner of land abutting the B3002 near Headley Down, Bordon, Hampshire ("the site"). The land was opposite Ludshott and Bramshott Commons, both of which were owned by the
B National Trust and constituted a designated site of special scientific interest. The site itself had been included by Hampshire County Council in a countryside heritage area. The local planning authority, East Hampshire District Council, alleged that the site was being used by the defendant, unlawfully and in breach of planning control, as a showman's winter quarters by placing on the land five residential caravans and 11 other vehicles or trailers.

C In 1981 temporary planning permission had been granted to a Mr Swoish for the placing of a mobile home on an area of land including the site. In 1985, following the expiry of that permission, the council served an enforcement notice requiring permanent discontinuance of the use of the land, including the site, "as a site for a mobile home". An appeal against that notice was dismissed. The defendant bought the site from Mr Swoish in September 1989. On January 8 1990 the council issued three summonses against the defendant
D in respect of his failure to comply with the 1985 notice. On January 10 the council issued a writ seeking, *inter alia*, an injunction restraining the defendant: (a) from occupying or residing in a mobile home or residential caravan on the site; and (b) from using the site for the storage of mobile homes, residential caravans, vehicle trailers and showground or funfair equipment. On January 22 the defendant applied for planning permission to use the site as a showman's winter quarters. On January 23 the writ was served and on the same day the three summonses came before the Alton justices and were
E adjourned to enable the defendant to obtain legal advice. On February 6 Mr Patrick Bennett QC (sitting as a deputy judge of the Queen's Bench Division) dismissed an application by the council for an interlocutory injunction, holding that, as a matter of construction, the 1985 notice was not wide enough to cover the placing of a mobile home on the site by the defendant. On March 6 the council gave notice of appeal against this decision. On March 7 the defendant's application for planning permission was refused and on March 23
F the council served three further enforcement notices alleging breaches of planning control arising from the present use of the site. Appeals by the defendant against the enforcement notices and the refusal of planning permission were due to be heard on December 4 1990.

Held The council's appeal was dismissed.

The application for an interlocutory injunction had to be determined in
G accordance with the principles set out in *American Cyanamid Co* v *Ethicon Ltd* [1975] AC 396. It was not in dispute that the defendant's activities were in breach of planning law but there was a serious issue to be tried as to whether it was a proper case for the grant of a permanent injunction: see p 11E. Damages would be a wholly inappropriate and inadequate remedy. However, the effect of enforcement by the criminal law had not been tested: see p 12B-C. The defendant had not done any permanent damage to the land. In the circumstances the balance of convenience was against the grant of an injunction having regard to the onset of the winter season, the difficulties the

defendant was likely to face in finding alternative quarters for the remainder A
of the season and the fact that the planning and enforcement notice appeals were only a few weeks away: see pp 12D-13A.*

Cases referred to in the judgments

American Cyanamid Co v *Ethicon Ltd* [1975] AC 396; [1975] 2 WLR 316; [1975] 1 All ER 504, HL

Stoke-on-Trent City Council v *B&Q (Retail) Ltd* [1984] AC 754; [1984] 2 WLR 929;
[1984] 2 All ER 332; (1984) 82 LGR 473, HL B

Appeal against decision of Mr Patrick Bennett QC

This was an appeal against a decision of Mr Patrick Bennett QC, sitting as a deputy judge of the Queen's Bench Division, on February 6 1990 whereby he had dismissed an application by the plaintiffs, East Hampshire District Council, for an interlocutory injunction.

James Findlay (instructed by Sharpe Pritchard, agents for the solicitor to East C
Hampshire District Council) appeared for the appellants.

Sebastian Head (instructed by Fox & Kent, of Chichester) appeared for the respondent, William Davies.

Cur adv vult

The following judgments were delivered

FOX LJ: This is an appeal by East Hampshire District Council ("the D
council") from an order of Mr Patrick Bennett QC, sitting as a deputy judge of the Queen's Bench Division, dismissing the council's application for an interlocutory injunction.

The case concerns land at Bowleswood Farm, Headley, Bordon, Hampshire. I will refer to it as "the site". It is within the council's district. The council bring the action under the powers conferred by section 222 of the E
Local Government Act 1972. The council are a local authority within the provisions of section 272 of that Act.

The site was formerly part of a larger agricultural estate and more recently has been divided into sub-plots of two or three acres each. The site is situate on the northern side of the B3002. Directly opposite the land, and on the south side of the B3002, are Ludshott and Bramshott Commons, both of which are owned by the National Trust and which constitute a designated site of special F
scientific interest covering about 370 ha and including extensive tracts of heathland. The site itself has been included in Hampshire County Council's Bramshott and Ludshott countryside heritage area. That designation is at present on a non-statutory basis. The public has access to the National Trust land. The site is not obscured from the public road except by a small hedge. There is also a 2-m high steel mesh fence bordering the site but it does not
prevent persons passing on the public road from seeing the whole extent of the G
site.

It is the council's case that part of the site is being used by the defendant unlawfully and in breach of the planning legislation as an area for a showman's winter quarters. That involved the placing of vehicles, including mobile homes, on the site and, say the council, constitutes a material change of
use. That change of use was unauthorised and a breach of the planning laws. H
There are upon the land, according to the evidence, five residential caravans or mobile homes, together with 11 assorted vehicles or trailers. The council's view is that these vehicles are a particularly offensive sight in this area of highly attractive countryside.

*Editor's note: On January 28 1991 the planning and enforcement notice appeals were allowed. The enforcement notices were quashed and personal planning permission was granted for use of the site as showmen's winter quarters until March 1993, subject to conditions: see (1991) 6 PAD 304.

A A planning authority (the council in the present case) may issue an enforcement notice in respect of a breach of planning control requiring the breach to be remedied. A person having an interest in the land may appeal against the notice and, if he does, the notice has no effect pending the final determination or withdrawal of the appeal (see section 88(10) of the Town and Country Planning Act 1971 [now section 175(4) of the Town and Country B Planning Act 1990]). An enforcement notice has effect against the owner of the land for the time being but, since appeals must be within four weeks from the notice, the owner for the time being may be out of time for appealing. If an effective enforcement notice is not complied with, the person in default is liable to a fine.

On October 5 1981 temporary planning permission was granted to a Mr Swoish for the placing of a mobile home on an area including the site. On C October 5 1984 the temporary planning permission expired.

On January 3 1985 an enforcement notice ("the 1985 enforcement notice") was served, taking effect on February 14 1985. This required that there be permanent discontinuance "of the use of the land as a site for a mobile home" and the removal of the mobile home from the land. Mr Swoish appealed against that enforcement notice. The appeal was dismissed on October 8 1985.

On September 22 1989 the defendant purchased from Mr Swoish an area of D land which included the site. Mr Swoish retained about 13 acres.

On November 7 1989 Mr Curtis, a development inspector in the planning department of the council, visited the site and met the defendant, who told him that he intended to use the land as a winter stop-over for his caravans and equipment. Mr Curtis gave the defendant a copy of the 1985 enforcement notice. On November 24 1989 Mr Curtis visited the site again and found on it five residential caravans or mobile homes and 11 other vehicles or trailers.

E On January 8 1990 the council issued three summonses against the defendant in respect of failure to comply with the 1985 enforcement notice.

On January 10 1990 the writ in this action was issued. The relief sought was (*inter alia*) an injunction restraining the defendant by himself, his servants or agents (a) from occupying or residing in a mobile home or residential caravan or otherwise howsoever on the site, and (b) from using the site for the storage F of mobile homes, residential caravans, vehicle trailers and showground or funfair equipment.

On January 22 1990 the defendant applied for planning permission to use the site as a showman's winter quarters (that is to say from October 1 to March 31 each year). On January 23 1990 the writ was served. On the same day, January 23, the three summonses came before the Alton magistrates and were adjourned for the defendant to obtain legal advice.

G On February 6 1990 the present application for an injunction came before Mr Patrick Bennett QC and was dismissed.

On February 20 1990 the three summonses again came before the Alton Bench and by consent were adjourned *sine die* pending the determination of the present appeal.

On March 6 1990 the council gave notice of appeal against the decision of H Mr Bennett.

On March 7 1990 the defendant's application for planning permission was refused.

On March 23 1990 the council served three further enforcement notices upon the defendant in respect of alleged breaches of planning control in consequence of the present user of the site.

The decision of the deputy judge turned largely on the effect of the 1985 enforcement notice. The background to that notice was the planning

permission granted to Mr Swoish in October 1981. That provided that for A
three years from the date of the permission or until the dwelling ceased to be occupied by the appellant (Mr Swoish), whichever should first happen, "the use hereby permitted" shall cease. The permitted use was the retention of a mobile home on the land. Schedule 1 to the enforcement notice states the premises to which the notice applies; those premises include the site. Schedule 2 states the alleged breach of conditions, ie that the use of the land as a site for a mobile home has not ceased after the expiration of the three-year period. B
Schedule 3 states the steps required to be taken to remedy the breach, ie "permanently discontinue the use of the land as a site for a mobile home and remove the mobile home from the land".

The deputy judge took the view that, as a matter of construction, the language of the 1985 enforcement notice was not wide enough to cover the placing of a mobile home on the land by the defendant. He therefore held that C
the notice does not necessarily have an application wide enough to include the admittedly unlawful activities (of the defendant) on the land. In these circumstances he refused an injunction.

Accepting that there may be questions of construction as to the ambit of the 1985 enforcement notice, I do not think it is appropriate to attempt to resolve them on an interlocutory application or that their resolution in favour of the defendant is sufficient to determine the present application. The 1985 notice D
and the three notices served this year are a basis for criminal proceedings (assuming them to be valid and enforceable according to their terms). As I have mentioned, the 1990 notices are under appeal and, therefore, of no effect pending the appeal. The 1985 notice, whatever its construction, adds little to the council's contention that the defendant's present activities in any event constitute unlawful development on the site and should be stopped until judgment in the action or further order. E

This is an application for an interlocutory injunction and must be determined in accordance with the *American Cyanamid* principles. As to those, there is, in my view, a serious issue to be tried in the action. It is not, I think, in dispute that the defendant's activities are in breach of the planning law. The issue to be tried is whether it is a proper case for the grant of a permanent injunction. The next question is whether damages at law would be F
an adequate remedy for the council. They would not. Damages are a wholly inappropriate and inadequate remedy for a local authority in these circumstances. Next, would the council's undertaking in damages, if an injunction were granted but the council lost at the trial, be an adequate remedy for the defendant? This gives rise to questions having regard to the position of travelling showmen. But, in any event, the matter under the *American Cyanamid* principles becomes a question of the balance of convenience (see G
American Cyanamid Co v *Ethicon Ltd* [1975] AC 396 at p 408E). I will return to that, but there is superimposed upon the question of the balance of convenience a further question as to the appropriateness of granting injunctions at all to a local authority in cases, such as the present, where Parliament has made special provisions for penalties for breach of the relevant statutes.

In *Stoke-on-Trent City Council* v *B&Q (Retail) Ltd* [1984] AC 754 (a case H
concerned with breaches of the Sunday trading legislation) Lord Templeman, whose speech was concurred in by the other members of the House, said at p 776 A-C:

> The right to invoke the assistance of the civil court in aid of the criminal law is a comparatively modern development. Where Parliament imposes a penalty for an offence, Parliament must consider the penalty is adequate and

A Parliament can increase the penalty if it proves to be inadequate. It follows that a local authority should be reluctant to seek and the court should be reluctant to grant an injunction which if disobeyed may involve the infringer in sanctions far more onerous than the penalty imposed for the offence. In *Gouriet* v *Union of Post Office Workers* [1978] AC 435 Lord Wilberforce said at p 481, that the right to invoke the assistance of civil courts in aid of the criminal law is "an exceptional power confined, in practice, to cases where an B offence is frequently repeated in disregard of a, usually, inadequate penalty . . . or to cases of emergency . . ." In my view there must certainly be something more than infringement before the assistance of civil proceedings can be invoked and accorded for the protection or promotion of the interests of the inhabitants of the area.

Lord Templeman went on to say that as a general rule a local authority should try the effect of criminal proceedings before seeking the assistance of the civil C courts but, in the *Stoke-on-Trent* case, the council were entitled to take the view that the traders would not be deterred by the maximum fine, which was substantially less than the profit which could be made from illegal Sunday trading.

The position in the present case is as follows. First, the effect of the enforcement of the criminal law has not been tested at all. No action can be taken by way of prosecution in respect of the 1990 enforcement notices D because their operation is stayed by the statute pending the appeals against them. So far as the 1985 enforcement notice is concerned, proceedings before the magistrates have been instituted by the council but have been adjourned, by consent of both sides, pending the outcome of the present appeal.

Second, it is a case in which the defendant has not done permanent or long-term damage to the land and, pending the outcome of the planning and E enforcement notice appeals, he offers an undertaking not to do such damage (though the precise terms of any such undertaking may require consideration).

Third, the planning and enforcement notice appeals will be heard in a few weeks, ie on December 4. The decision will not follow immediately but will be given within a period estimated at between one and three months. I should mention that though the matter was heard before the deputy judge in February F it did not come before the Court of Appeal until the Michaelmas Term. Apparently there was some misunderstanding as to the intention of the council to pursue the appeal and it was struck out of the list.

Fourth, the defendant's evidence is that he lives in a mobile home on the site and has nowhere else where he can store his vehicles and equipment. The legislation, I should observe, shows a degree of protectiveness in relation to a resident in a mobile home in that section 90(2)(*b*) of the Town and Country G Planning Act 1971 [section 183(3)(b) of the 1990 Act] provides that a stop notice shall not prohibit the use of land as a site for a caravan occupied by any person as his only or main residence.

Moreover, the showman's winter season is now upon us and it is difficult for showmen to find winter quarters. To turn the defendant out now when the planning appeals are still pending could, if planning permission were granted H on the appeals, cause damage to the defendant which might be disproportionate to the temporary protection of any public interest.

Looking at the whole matter, I think that the balance of convenience is in favour of refusing an injunction. I am not criticising the council for starting the proceedings; they are properly concerned for the public interest in the amenities of the countryside. Nor am I saying that the defendant has a licence to remain indefinitely. I am saying only that, having regard to the onset of the winter season, the difficulties likely to face the defendant in finding quarters

for the remainder of the season and the fact that the planning appeals are only A
a few weeks away tilt the balance of convenience, in a case where the criminal penalties have not been applied, in favour of the defendant at this stage.

I would, therefore, dismiss the appeal.

BUTLER-SLOSS LJ: I agree.

Appeal dismissed with costs in the Court of Appeal; costs order below to stand. B

Barratt Leeds Ltd *v* Secretary of State for the Environment and another

QUEEN'S BENCH DIVISION C

Mr Roy Vandermeer QC (sitting as a deputy judge of the Queen's Bench Division)

December 7 1990

Circular 16/87 — Test of agricultural land quality — Whether inspector misled applicant about agricultural land issue — Whether inspector applied test properly — Whether agricultural consideration properly sole issue D

By a decision letter dated April 25 1990 the Secretary of State for the Environment, in accepting his inspector's reasons and conclusions, dismissed the applicants' appeal against the decision of the second respondents, Rotherham Metropolitan Borough Council, to refuse planning permission for residential development on land at Sandy Lane, Bramley, Rotherham. There had been an earlier appeal in respect of substantially the same site, E which was dismissed in July 1986, the question of agriculture having been central to the issues then considered by the inspector. During the course of the 1989 inquiry in respect of a further application for planning permission a third party raised the question of agricultural land; reference was made to Circular 16/87, para 4. The inspector asked questions about agricultural land classification at the inquiry. The applicants challenged the Secretary of State's decision on the grounds that the inspector had recommended the dismissal of F their appeal solely by reference to an issue, namely agriculture, which had not been an issue at the inquiry. They had been misled by the inspector about the agricultural question. Further, the inspector had made no findings as to whether the land was the best and most versatile land in relation to agriculture and treated the grading of the land as an overriding consideration when it was only one of the relevant factors to be taken into account. He had failed to have proper regard to the policies in Circular 16/87. G

Held The application was allowed.

The inspector did not mislead the applicants about the agricultural land question; they misread the situation and the significance of agricultural considerations: see p 16D. Under the agricultural land classification, the land in question was class 3A. It was not possible to decide if this was the best and most versatile without determining if there was any class 1 or 2 in the area. Accordingly, the inspector had failed to apply the test in Circular 16/87 of H determining the "best and most versatile" agricultural land: see p 19B. There was no evidence before the inspector to give additional weight to the agricultural considerations. It was incumbent to put in the balance the practical considerations of whether and to what extent the land was in fact a useful farming unit: see p 19.

Case referred to in the judgment

Simplex G E (Holdings) Ltd v *Secretary of State for the Environment* [1988] 3 PLR 25

Application under section 245 of the Town and Country Planning Act 1971
This was an application under section 245 of the Town and Country Planning Act 1971 (section 288 of the Town and Country Planning Act 1990) to quash a decision of the first respondent, the Secretary of State for the Environment, by his inspector who had dismissed an appeal by the applicants against a refusal by the second respondents, Rotherham Metropolitan Borough Council, to grant outline planning permission.

Richard Phillips QC and Andrew Tait (instructed by Walker Morris Scott Turnbull, of Leeds) appeared for the applicants, Barratt Leeds Ltd.

John Steel (instructed by the Treasury Solicitor) appeared for the first respondent, the Secretary of State for the Environment.

The second respondents, Rotherham Metropolitan Borough Council, did not appear and were not represented.

The following judgment was delivered.

MR ROY VANDERMEER QC: This is an application by Barratt Leeds Ltd under section 245 of the Town and Country Planning Act 1971, now section 288 of the Town and Country Planning Act 1990, to quash a decision of the Secretary of State for the Environment made by a decision letter dated April 25 1990. In that decision letter, the Secretary of State accepted the recommendation of his inspector, Mr J M Steers, that the appeal should be dismissed. The appeal had been against the refusal of Rotherham Metropolitan Borough Council to grant outline planning permission for residential development on land at Sandy Lane, Bramley, Rotherham.

In accepting his inspector's recommendations, the Secretary of State added nothing to the reasoning of the inspector and, as he was perfectly and properly entitled to do, simply indicated that the recommendations were accepted for the reasons given by the inspector.

The consequence of that is simply this, that in questioning the efficacy of that decision, or in looking at the complaints that are now made, one has to look essentially, indeed only, at the inspector's report.

Before considering the arguments that have been put before me, I think it is important in the circumstances of this particular case to note that a previous appeal had been dismissed on July 30 1986 in relation to a site which while not identical was substantially the same as the present appeal site. That appeal also related to a proposed residential development.

In that case, the question of agriculture was central to the issues considered by the inspector and, in due time, by the Secretary of State. Although there was a reference to grade 2 land in some of the documentation, the Secretary of State dealt with the matter, when considering his inspector's report, on the basis that the land was grade 3A in terms of agricultural land quality.

The inspector had accepted the evidence concerning that land but recommended that planning permission nevertheless be granted. In part at least, that was because of his assessment of the practical opportunity to use that agricultural land, because of matters such as the fragmentation of a holding, the likelihood of vandalism and trespass and so forth. These matters led the inspector to describe the site as seriously constrained in agricultural terms.

The Secretary of State took a rather more simplistic view, in the sense that he approached the agricultural issue on a policy basis. In the light of what was clearly the policy approach then current, as appears in para 6 of the decision letter in that case, he concluded that land of better quality should be protected wherever possible and, in short, decided that the all-important thing was the grading of the agricultural land.

I should also record that while it is agreed that certain aspects of the policy

A have changed, Circular 22/80 — *Development Control: Policy and Practice* — to which the Secretary of State made reference on that occasion, remains extant policy guidance.

So far as the appeal that led to this matter coming to this court is concerned, by the time that that appeal took place the council themselves had an eye, if I can put it that way, so as to avoid questions of the particular status of their views on the development of this land for high technological purposes. They B did not raise an agricultural objection. To that extent, the agricultural matter was not central, at least at the outset, to the 1989 appeal.

However, it was plainly necessary, and certainly advisable, in the light of the previous decision, for the appellants to say something about the previous decision and the agricultural issue there raised. In fact they did, as I understand it, say something about it, because they asserted that there had C been changes of circumstances both in relation to the proposals to develop the site and in relation to Government policy with regard to agricultural land.

The importance of that is that, whatever else, the appellants patently recognised that the agricultural issue could not simply be brushed aside, albeit that it was not central. In my judgment, they patently must have had very much in their minds the fact that on the first occasion the appeal failed in relation to the agricultural issue substantially upon the basis of the D agricultural land classification and the quality of the land in that regard.

During the course of the inquiry, the question of agricultural land was raised by a third party, namely the Council for the Preservation of Rural England ("CPRE"). Their submissions were recorded by the inspector at paras 258 to 275 of his report. They drew attention to the agricultural land quality and at para 258 referred to certain percentages which, as I have been informed in this court, do not strictly relate to the appeal site but relate to a E somewhat larger area.

The agricultural point that they took was essentially brought together in para 259 of the report, where there was an objection specifically based upon Circular 16/87 — *Development involving Agricultural Land* — which relied upon the agricultural land quality and classification and asserted, as I see it quite correctly, that para 4 of Circular 16/87 stated that the best and most F versatile land has a special importance and should not be built upon unless there is no other suitable land available. They then referred to the additional weight that they said needed to be given to the agricultural considerations in such circumstances and went on to assert that this site fell within the categories to which para 4 related. They did so by reference to Annex A of the circular, about which I will have something to say later. That was plainly their submission. It was a submission that turned upon the agricultural land G quality.

The inspector himself asked some questions upon agricultural matters. As I both read it and understand from the submissions made in this court, the inspector's questions in fact related to the agricultural land classification and sought to ascertain whether this land was in fact classified in certain grades.

Mr Williamson, [the solicitor] who appeared on behalf of the appellants [at the inquiry], was somewhat concerned about the inspector's questions and H sought to ascertain a number of matters from him. The approach that Mr Williamson adopted to ascertain the inspector's position is set out in a number of affidavits. Mr Williamson has produced two affidavits. There is also an affidavit from Mr Carter, who gave planning evidence on behalf of the appellants, and there is also an affidavit from the inspector.

In essence, they come down, as both Mr Phillips and Mr Steel submitted, to the fact that the inspector made a claim that he was only asking questions

A about land classification. He did not answer, in terms, the question put to him as to what weight he attached to the point. I must confess that it might have been rather difficult for an inspector, certainly when the Secretary of State was making the decision, to give very useful guidance upon that point. Even if he had indicated his own feelings, it would not have been an end of the matter.

In the light of that indication from the inspector, that it was a matter of land classification and only that he was seeking to further investigate, Mr B Williamson initially sought the inspector's guidance upon whether the evidence of the farmer or evidence given at the earlier appeal would be of assistance. I need not go into the details of the exchanges; suffice it to say that the inspector indicated that he did not think he would be much helped, certainly on the point that he was investigating, by such material. Mr Williamson nevertheless sought to get proofs of evidence that had been given C at previous inquiries, as well as other information, but was unable to do so by the time the inquiry reached its termination. He did not seek an adjournment.

It is said on behalf of the appellants that the inspector's intervention, taken together with his answers and reaction to Mr Williamson, was so put as to mislead the appellants as to the weight that was being attached to the agricultural issue. It is said by Mr Phillips that they were misled into the belief that nothing very much could or would turn upon the agricultural issue. That D that was the state of mind that the appellants, through their advocate, were in, I have no doubt.

While I have a measure of sympathy for the appellants, in my judgment this is not a case of the inspector misleading the appellants. It is patently a case of the appellants misreading the situation. I do not seek to criticise Mr Williamson. It is much easier with the benefit of hindsight to look at and analyse what had gone on and draw conclusions and say: "Well, that is what E one should have realised".

The reason that I recited the history of the earlier hearing and the Secretary of State's subsequent decision was that, certainly with the benefit of hindsight, it seemed patently obvious to me that the Secretary of State had primarily acted on land classification on the earlier occasion. That was basically the approach of the CPRE on the appeal with which I am concerned. The inspector was asking questions which he said only went to land F classification, but which had itself been at the heart of the previous refusal.

Again with the benefit of hindsight, it seems to me that the warning lights were there, that agriculture was looming its head, as inevitably it did, once the CPRE objected. It must, in my judgment, when somebody raises an objection, be primarily and fundamentally, if not only, for the appellants to determine how to meet that particular objection.

G In the event, of course, it is apparent that the inspector was, and it is indeed part of Mr Phillips' complaint, interested in land classification, because Mr Phillips tells me, with some force, that the inspector ultimately decided this case very much upon the basis — I would not say exclusively, but very much upon the basis — of the land classification. For those reasons I reject Mr Phillips' submission, as very helpfully set out in the first of his list of issues, and in effect reject those matters raised in the notice of motion at paras 1 to 3 H inclusive*.

---

*Editor's note: Paras 1 to 3 of the notice of motion read as follows:

"1. The First Respondent, following his Inspector's conclusions, dismissed the Appellants' appeal and solely by reference to an issue, namely agriculture, which was not an issue at the Inquiry between the Appellants and the Second Respondents and upon which the Appellants were not given an adequate opportunity of addressing evidence and/or of making full submissions.

2. The Inspector misled the Appellants at the Inquiry as to the relevance and significance of the agricultural issue and thereby failed to give the Appellants an adequate opportunity of addressing evidence and/or of making full submissions.

3. In the circumstances the First Respondent acted unfairly and was in breach of the rules of natural justice."

I now turn to the matter raised in para 4(1)(a) of the notice of motion and dealt with in Mr Phillips' list of issues as the second of them. In my judgment, this is the most important issue of those that are left.

What is said in the notice of motion at para 4(1)(a) is that the inspector made no finding and no conclusions as to whether the appeal site comprised the best and most versatile land and alternatively he was wrong to conclude that the appeal site did constitute the best and the most versatile land.

Mr Phillips developed that matter by indicating that the inspector's consideration of the question of whether it was the best and most versatile was incorrect, and there is indeed some doubt as to whether that was the conclusion the inspector reached.

I have already indicated the way in which the matter was dealt with previously before the Secretary of State and the way in which the CPRE approached it and the way in which the inspector approached it in terms of concentrating to a considerable extent on agricultural land quality.

The matter is dealt with in a number of places, but I turn first to para 288 of the inspector's report. In that paragraph the inspector, while identifying the site as, at least for the time being, being part of the countryside, described the land as predominantly classified as grade 3A with some grade 2 land adjacent to the motorway. He referred to the land classification evidence of the CPRE and the lack of challenge to that aspect of the evidence.

He later, in para 290, referred to the land only in the context of grade 3A. The explanation for that is that such grade 2 as existed was a very small part of the whole and, indeed, as the inspector himself found, was affected by overhead power lines. I think it is quite clear that, ultimately, the inspector reached his agricultural judgment upon the large part of the site and its grade 3A quality, and I do not undertand Mr Steel to have suggested otherwise.

In para 290 the inspector quite properly recorded that, since the previous appeal, there had been the new Circular 16/87. He recorded that the appellants had sought to counter the earlier decision as to the need to preserve grade 3 land by referring to the size of the site which, therefore, did not need to be put before the Ministry of Agriculture, Fisheries and Food for consultation purposes.

I am not sure that that fully records the point put by the appellants, because they did, as recorded, indicate that there had been a change in Government attitude, although they certainly did make a point about size. Be that as it may, the real question is whether the inspector got it right in the light of the circulars to which he was properly referring.

Having recorded the appellants' argument, the inspector went on, in para 290, to say that albeit that it was no longer necessary to refer the matter to the Ministry of Agriculture, Fisheries and Food, that, and I use his words, "still does not overcome the fact that" — and then he cited from Circular 16/87 — "additional weight needs to be given to the agricultural factors".

I find some difficulty in establishing whether the inspector had specifically come to the conclusion that this land was, in terms of Circular 16/87, the best agricultural land. It may be that one could, in other circumstances, if that were the only issue, reasonably move to an inference about that matter. What is clear is that if it were not the best agricultural land, the additional weight to which the inspector referred would not, in terms of the Government guidance, apply.

It is necessary for me to refer to para 4 of Circular 16/87. Before doing so, however, I note that para 3, as indeed does, in a sense, Circular 22/80, contains guidance that makes it plain that one should not be liberal with the use of agricultural land, although para 3 makes it plain that other factors need to be considered in the general issue of agricultural land.

A I think it may be useful if I were to set out para 4 in full. The Government guidance appears thus:

> In assessing these various factors, while the agricultural quality of the land will often be a relevant consideration, it also has to be borne in mind that the demand for agricultural products fluctuates over time and, once land is developed, return to agricultural use is seldom practicable. The best and most versatile land has a special importance and should not be built on unless there
> B is no other site suitable for the particular purpose (advice on the relative quality of agricultural land is given in Annex B). This land is a national resource for the longer term and should in general be protected from irreversible development. In such cases, therefore, additional weight needs to be given to the agricultural factor.

It seems to me to be plain beyond peradventure, and I do not understand Mr
C Steel to be challenging it, that it is only in the case of the best and most versatile, guidance as to which is given in Annex B, that that additional weight emerges.

At this point, I would reiterate that the CPRE, in making their point, had referred to Annex A and not to Annex B. In my judgment, Annex A does not tell us what is the best in the context of para 4 but gives guidance which is described in the heading to Annex A as relating to procedural arrangements
D and, in certain circumstances, subject primarily, that is, to the size of the proposed development and related areas, causes land of grade 1, 2 and 3A to be referred to the ministry.

That is not something that emerges from the annex which para 4 guides one to. It seems to me, whatever judgment one makes as to what is the best in the context of para 4, that is not affected by Annex A, which is, if you like, a precautionary step to allow the ministry in any given case to consider whether,
E in the circumstances of that case, grade 3A is to be regarded as the best and how the matter should be approached.

It is, in my view, of some significance that the inspector in addressing the question of additional weight at para 290 referred to para 259, whereas the CPRE themselves had referred only to Annex A and not to Annex B.

If I turn to the guidance at Annex B, there is a little difficulty in that
F guidance itself in that in para 1, grade 1 is described as the best. At that stage, the words "and most versatile" do not appear. In para 3, however, the guidance is that land graded 1 and 2 is the highest quality and has few limitations in agricultural use, and then it gives the percentage that it represents. It is then described as "versatile" and most efficient in response to inputs. Grade 3 land is dealt with at para 4 and is described not as the most versatile but as having moderate physical limitations. The paragraph goes on
G to indicate that there is a wide range within grade 3 and describes 3A as having "comparatively few limitations", and it then talks of the qualities of the other land within the subclassification of grade 3. It does not refer to 3A land as "the best and most versatile".

Against that background, it seems to me that the fundamental paragraph is para 6 of Annex B. We are there told that in many parts of the country, where there is no grade 1 or 2 land, grade 3A will represent the best and most
H versatile. It seems to me, with respect to Mr Steel, that to read that as other than meaning that there will be places where grade 3A will not represent the best and most versatile is to strain the meaning of words unmercifully.

It seems to me, therefore, that having established the land as grade 3A, it was impossible to consider whether, in the circumstances of this case, it represented the best agricultural land attracting additional weight to the agricultural issue without having at least some information or evidence, certainly something, which would indicate whether this was a part of the country or an area where the provision of grade 1 or grade 2 land was of such an order, or whatever, as to lead it to be judged that grade 3A would, in those circumstances, represent the best and most versatile. This must more particularly be the case in the light of the language of para 4 of Annex B to which I have referred.

It follows, in my judgment, that the inspector had either failed to address properly the circular and the test of what the best was or concluded, without any basis, that it was in fact the best. What is plain, in my judgment, is that there was nothing before the inspector, and no basis in Circular 16/87, that could properly lead him to approach this case on the basis that additional weight should be given to the agricultural issue in the light of the circular.

For those reasons, I find that Mr Phillips' complaint, set out in his ground 4(1)(a) of the notice of motion, is valid and, for that reason, this decision notice must be quashed.

I no longer need to address, as at one time in the light of Mr Steel's argument I might have had so to do, the question of whether the decision should be upheld on the basis that the inspector had also mentioned matters relating to open countryside. In the light of the decision in *Simplex G E (Holdings) Ltd* v *Secretary of State for the Environment* [1988] 3 PLR 25, Mr Steel has stepped back from that argument and, during the course of argument, accepted that if the complaint that Mr Phillips made with regard to para 4(1)(a) was substantiated, then there was no basis on which this decision letter could be saved.

It perhaps is unnecessary for me to say every much about the countryside point in those circumstances, but I will just permit myself to say that, for my part, the validity and strength of the countryside point must be approached with some measure of reservation against the light of the fact that the inspector, as is plain from the language that he used in para 291, where he talks about the need to protect the countryside for its own sake, was relying upon the words of Circular 16/87 in circumstances where there was, as I understand it, no specific local policy that this land should be kept open but, on the contrary there were emerging policies that it should be developed. Be that as it may, that matter adds nothing in terms of the issues that are before me.

In the light of what I have said, it follows that the other points made by Mr Phillips can be dealt with pretty shortly and I will say but a word about them.

The third point that Mr Phillips takes, relating to para 4(1)(b) of the notice of motion, suggests that the inspector was incorrect to treat the agricultural grading as an overriding consideration to the exclusion of all other factors. I think I can confine myself to saying that having reached the conclusion that I have with regard to the inspector's approach to the question of extra weight, the very basis upon which the inspector moved to treating agriculture as an overriding consideration was flawed. Approached on the proper basis, and even on the basis that this was the best agricultural land, it is plain that the matter should not stop there. In my judgment, it was incumbent to put in the balance the practical considerations of whether and to what extent this land was in fact a useful farming asset.

I turn, then, to the fourth issue advanced by Mr Phillips, relating to the notice of motion in para 4(1)(c), whether the inspector paid any regard to the change brought about in Government policy. I need spend little time on this. There is little indication, as Mr Steel in fact asserted, as to whether the inspector identified a change. I suppose in this case the question of whether there was a change of policy was of some significance, since the inspector at least gave the impression, as one reads the letter, of simply following the

A Secretary of State's line of some years before without paying too much regard to change.

That having been said, had the inspector properly applied Government guidance as set out in Circulars 16/87 and 22/80, it would not perhaps have been of any consequence. I do not think I need to make a particular finding upon whether he paid proper regard to the question of whether there had been a policy change. The real point is that he got Circular 16/87 wrong.

B Turning to the fifth issue raised by Mr Phillips, whether the inspector had proper regard to the adopted policy of the second respondents in regard to the future use of the appeal site, I think the answer must be that he did not. His approach to this issue was interlinked to his view on agricultural land and his conclusion that additional weight needed to be given to the agricultural aspect. This in turn led him to approach all other matters upon the basis that they C needed to be overriding considerations when weighed against the agricultural considerations. Again his conclusions on the agricultural issue tainted his approach to other matters.

Clearly, then, he did not properly regard the local plan, because he put it in a balance where the scales were weighted wrongly on the other side. In those circumstances I do not need to consider the question of whether he properly understood the precise process that was going on in relation to local policies D or, indeed, whether Mr Phillips is right in suggesting that the inspector had made a factual error as to the way in which that site was being brought forward. Nothing in the circumstances of the way that I have dealt with this matter seems to turn upon that issue. Certainly, the question of the emerging policy, and of the council's acceptance of the site as a development site was a relevant consideration, but I need say no more about that matter.

For those reasons, this decision notice will be quashed.

E *Application allowed with costs.*

Barber v Secretary of State for the Environment and another

F

QUEEN'S BENCH DIVISION

Mr David Widdicombe QC (sitting as a deputy judge of the Queen's Bench Division)

December 21 1990

Planning gain — Application for residential development and proposed G bypass — Applicant not owning land for bypass but agreeing with highway authority to pay most of bypass cost if permission for residential development granted — Inspector finding residential development unacceptable — Whether Circular 22/83 applied where bypass part of application

The applicant, Arthur Barber, appealed against the refusal by Horsham District Council of an outline application for a residential development on the H outskirts of Ashington, West Sussex, together with a proposed bypass on the east of the village. He did not own the land for the bypass but he had entered into two agreements with the county council as highway authority whereby, if planning permission for the residential development was granted, he would pay the council £4.5m, nearly the full cost of the bypass. At the inquiry part of his case was that the bypass was needed to accommodate traffic generated by the residential development, so that the two elements of the application were directly related. The inspector dismissed the appeal, concluding that: (1) the

residential development was unacceptable from a planning point of view; (2) A
the benefit of the bypass was largely extraneous to the residential development; (3) the principles laid down in Circular 22/83 applied even though the bypass was part of the application; (4) the bypass failed the tests in the circular; and (5) the harmful aspects of the residential development were therefore not outweighed by the benefits to the public of the bypass. The applicant applied to the High Court to quash the inspector's decision on the ground, *inter alia*, that the inspector was wrong to deal with the application as B
if Circular 22/83 applied to it and that the proper balancing exercise was never carried out.

Held The application was dismissed.

Circular 22/83 does not expressly apply where the planning gain arises from development included in the application for permission, but the inspector had recognised this in his decision: see pp 24H-25A. The inspector made a finding of fact that the bypass was extraneous to the residential development and thus C
he was entitled to apply the criteria in the circular as his guide: see p 26F-G. Further, any planning gain lay not in the grant of permission for the bypass but in advancing its timetable and paying for it: these benefits were secured not by the inclusion of the bypass in the application but by the collateral agreements with the highway authority: see pp 26H-27A.

Cases referred to in the judgment D

R v *Westminster City Council, ex parte Monahan* [1990] 1 QB 87; [1989]3 WLR 408; [1989] 2 All ER 74; [1989] 1 PLR 36, CA

Safeway Properties Ltd v *Secretary of State for the Environment* [1990] 3 PLR 87; (1990) 61 P&CR 280; [1990] JPL 759

Application under section 245 of the Town and Country Planning Act 1971

This was an application under section 245 of the Town and Country Planning Act 1971 (section 288 of the Town and Country Planning Act 1990) to quash a decision E
of the first respondent, the Secretary of State for the Environment, by his inspector, who had dismissed an appeal by the applicant, Arthur Barber, against the decision of the second respondents, Horsham District Council, to refuse an application for outline planning permission for residential development at Oast House Farm, Ashington, West Sussex, together with the construction of a bypass to the east of the village.

Clive Newberry (instructed by Camerons, agents for A R Drummond & Co, of F
Tadworth, Surrey) appeared for the applicant, Arthur Barber.

Michael Burrell (instructed by the Treasury Solicitor) appeared for the first respondent, the Secretary of State for the Environment.

David Keene QC (instructed by the solicitor to Horsham District Council) appeared for the second respondents.

The following judgment was delivered. G

MR DAVID WIDDICOMBE QC: This is an application under section 245 of the Town and Country Planning Act 1971 [section 288 of the Town and Country Planning Act 1990] to quash the decision of an inspector appointed by the Secretary of State for the Environment whereby he dismissed an appeal against the refusal of planning permission for residential land at Oast House Farm, Ashington, West Sussex. H

The application for planning permission, which was in outline, included two main elements: a proposal to build not more than 95 dwellings on the northern outskirts of the village of Ashington and a proposal to construct a bypass to the east of the village. The land for the bypass was not in the applicant's ownership but he entered into two agreements with the county council, as highway authority: one, under section 52 of the Town and Country Planning Act 1971 [section 106 of the 1990 Act], conditional on the grant of planning permission for the residential development, not to commence that development until work on the bypass had been started and not to occupy or permit to be occupied more than 50 dwellings until the bypass was open to traffic; and the other, in the form of heads of agreement, under section 278 of the Highways Act 1980 to be entered into if planning permission for the residential development was granted, whereby the council covenanted to use their best endeavours to commence construction of the bypass in 1993 or as soon as possible thereafter and Mr Barber covenanted to pay the council the sum of £4.5m. It was envisaged that the county council would acquire the land for the bypass, compulsorily if necessary. I was told that £4.5m is virtually the full cost of the bypass.

Thus Mr Barber was saying, in effect: "If I get planning permission for the houses, I will pay for the bypass". It was part of his case at the inquiry that the bypass was needed to accommodate the traffic generated by the residential development and that the two elements of the application were therefore directly related.

After a local inquiry the inspector gave his decision on the appeal by letter dated July 10 1990. In summary, he said that: (1) the residential development was unacceptable from a planning point of view; (2) the benefit of the bypass was largely extraneous to the residential development; (3) the principles laid down in Circular 22/83 on "planning gain" applied even though the bypass was part of the application; (4) the bypass failed the tests in the circular; and (5) the harmful aspects of the residential development were therefore not outweighed by the benefits to the public of the bypass.

Paras 2 and 3 of his decision letter were as follows:

> 2. From all that I have seen and heard I consider that the main issues in this appeal are whether, in the light of approved planning policies, the proposed development would be an unacceptable extension of Ashington into the surrounding countryside, or have an unacceptable impact upon the character of the village, and if so, whether benefits to the public at large arising from the advancement of the Ashington Bypass would be such as to outweigh the planning disadvantages of the proposal.
>
> 3. The essence of your case was that the appeal proposal was acceptable on its own merits. It would also bring with it substantial benefits to the community in bringing forward the Ashington Bypass, to which your client had agreed to contribute a fixed sum of £4.5m, subject to the grant of planning permission, and the bypass being constructed at the earliest possible date.

In paras 5 to 18 the inspector considered the residential element of the application. I need only read para 18 which summarises his conclusions on that:

> To summarise, the appeal proposal would cause demonstrable harm to countryside which is of the quality which should be protected for its own sake: it would harm the character of the village by perpetuating the type of development which has contributed little to its present character, and by eroding its rural setting. It would run counter to the approved policies which have as their objective the protection of the environment. It would be followed by further significant amounts of development which would also be harmful. There would have to be strong planning benefits from the proposal to outweigh this degree of harm.

Mr Newberry, for the applicant, criticised this paragraph and some of the others dealing with the residential development because they referred to the "appeal proposal" as if the appeal related only to the residential element of the application. However, I do not think the inspector was ignoring the fact

that the bypass was included in the application. The second half of the decision A
is devoted to that subject. The references to the "appeal proposal" in para 18 and elsewhere are not misleading; in context they are simply references to that part of the proposal which the inspector was then considering, namely the residential part.

Paras 19 to 24 are concerned with the bypass and the subject of planning gain. It is this part of the decision which is directly challenged and so I will read it all. B

> 19. The construction of the Ashington Bypass is urgently required. It has been needed for many years, and it is quite apparent that it is needed not just in order to provide relief for the people who live in Ashington, but that the present road is quite inadequate for the traffic using it as part of the national network, and that many accidents could be avoided by the construction of the bypass. No other development is likely to fund the proposal as completely as your client proposes. His proposal would be a contribution to public funds, C
> and in my view it would be of the nature of a benefit extending beyond the development for which planning permission has been sought. I recognise that permission is sought for the bypass as part of the application, but I see no reason why the principles of Circular 22/83 should not apply.
>
> 20. The circular sets clear tests as policy. First, it is stated that a wholly unacceptable development should not be permitted just because of extraneous benefits offered by the developer. In my view the development D
> comes close to being wholly unacceptable in terms of the approved policies. The remaining tests set bear upon whether the benefit is extraneous. The bypass is needed to enable the development to go ahead, in that there are highway objections to increasing traffic at the junction of the A24 and Billingshurst Road. However, these objections could be overcome by far less substantial alterations than are entailed in the bypass, and thus the extent of what is proposed is not reasonably related in scale to the proposed E
> development. For these reasons it is not so directly related to the development that the proposal should not be permitted without it. Furthermore, it seems to me that the funding of the bypass is more appropriately found from national and local taxation. In this I agree with the Inspector in the 1988 appeal who could see no good reason why such urgently needed public works should have to be substantially privately funded.
>
> 21. You, however, took the view that the fact that a financial benefit would accrue was in itself a material consideration; a contention with which I agree F
> as far as consonant with government policy and its interpretations. In your view it was unrealistic to consider that the County Council would build the road without a substantial contribution from developers. This was evidenced by the statements of the County Surveyor at meetings with your client; the history of the project; the statement in the current TPP that an acceptable development package may materialise and enable the scheme to be brought forward; and the agreement of both Council and Parish Council witnesses in G
> cross-examination that desirable as they might find it for the scheme to be wholly funded from the public purse, the County Council would not do so whilst there was the prospect of a developer funding the scheme. That being the case, the development had clear public benefits which should go into the balance.
>
> 22. I refer again to Circular 22/83. An authority is not entitled to treat an applicant's need for permission to obtain some extraneous benefit for the benefit of ratepayers at large. Yet this appears to be the case in this instance. H
> Further, it is clear from the policies of the County Council, as expressed in the Structure Plan, and to which I give greater weight than the TPP, that there is no justification for expecting a substantial contribution, since only small scale development is envisaged in Ashington. Finally, Policy M9 gives a clear priority to road schemes on the A24 and A264. Of these, only the Ashington Bypass and Hookland Improvement remain to be constructed, and thus must have the highest priority in terms of approved policies.

> 23. The Ashington Bypass and the Hookland Improvement are now treated separately. The Improvement is due to start in 1991/92, whilst the Ashington Bypass is not programmed, although it has been resolved to commence land purchase procedures. At my request an engineer attended the inquiry to explain the priority ranking system of the TPP. This is set out as four criteria. By comparison, the Ashington bypass scores as highly or better than the Hookland Improvement on all four criteria. The schemes are of similar cost, and the capital budget is adequate to fund both within a reasonable time.
>
> 24. I therefore conclude that there is no justification in either the approved policies or the transport policies of the County Council for your view that the bypass would not be constructed within a reasonable time without the contribution proposed by your client. That is no criticism of your client for making the offer, but since there is no reason why the bypass should not have been brought into the programme for a start at the earliest possible date, no benefits in bringing forward the scheme can be attributed to his proposal in order to weigh against the disadvantages of the development.

Mr Newberry criticised this part of the decision on the ground that it ignored the fact that the bypass was part of the application and dealt with it as if Circular 22/83 on planning gain applied to it. He did not say that Circular 22/83 was irrelevant, but its application in this case should take into account that the bypass was an integral part of the application. He criticised para 20 because it failed to treat the bypass as part of the application and questioned the relevance of the reference to funding from national and local taxation. On para 21 he said that the inspector should have considered whether the residential development was acceptable on the assumption that the bypass was in position. He criticised paras 23 and 24 on the ground that the inspector was inventing his own priority for the roads and his conclusion in the first sentence of para 24 was unjustified on the evidence.

He referred to *R* v *Westminster City Council, ex parte Monahan* [1990] 1 QB 87* ("the *Covent Garden* case"). He said what the inspector should have done was to weigh the benefit of the bypass against the planning objections to the residential development and that balancing exercise was never carried out.

Mr Burrell, for the Secretary of State, said that the inclusion of the bypass in the application was merely a device to try to avoid Circular 22/83 on planning gain. He defended the inspector's decision and said that the approach that he had adopted, namely to consider the residential development first and then the bypass, was sound. He maintained that Circular 22/83 did apply to a case where the application itself contained the benefit and its principles had been correctly applied in this case. He referred to the *Covent Garden* case and to *Safeway Properties Ltd* v *Secretary of State for the Environment* [1990] JPL 759†, a decision of Otton J, which he told me was under appeal§.

Mr Keene, for the local planning authority, Horsham District Council, also supported the decision. He pointed out that the grant of planning permission for the bypass was of no benefit in itself; any benefit lay in the bringing forward of the bypass in point of time and the making of a financial contribution to the cost of construction. The inspector was right to apply the principles of the circular. He had found as a fact that the two elements of the proposal were unrelated and the inclusion of the bypass in the planning application did not overcome that.

The nub of Mr Newberry's case is that the inclusion of the bypass in the application makes a difference to the approach which should be adopted on the question of planning gain. I think he is right that Circular 22/83 does not

*Editor's note: Also reported at [1989] 1 PLR 36.
†Editor's note: Also reported at [1990] 3 PLR 87.
§Editor's note: The appeal was allowed by the Court of Appeal on June 14 1991.

expressly apply to cases where the planning gain arises from development A included in the application for permission. I think the inspector recognised that when he said in para 19:

> . . . I see no reason why the principles of Circular 22/83 should not apply.

The relevant parts of Circular 22/83 are as follows:

> **Definition** B
>
> 2. "Planning gain" is a term which has come to be applied whenever, in connection with a grant of planning permission, a local planning authority seeks to impose on a developer an obligation to carry out works not included in the development for which permission has been sought or to make some payment or confer some extraneous right or benefit in return for permitting development to take place. As such, it is distinct from any alterations or modifications which the planning authority may properly seek to secure to the development that is the subject of the planning application — such as C changes intending to reduce the scale or intensity of the proposed development, or to improve its layout or its impact on the local environment. In the case of "planning gain" the obligation sometime arises from the terms in which the development is permitted, eg from a condition of the planning permission, (conditions are dealt with in Ministry of Housing and Local Government and Welsh Office Circular 5/68 which is currently being revised) and sometimes from an agreement made in association with it. In D some cases the developer may offer some such works or payment in applying for planning permission or in the course of subsequent negotiations: the advice in this circular is relevant to those circumstances as well as to cases where the authority seeks to impose such obligations. This circular is not concerned with cases where the authority is disposing of land which it owns and where the terms and conditions on which it is prepared to sell are matters for negotiation with prospective purchasers. Nor is it concerned with matters arising from other legislation, eg the requisitioning of the provision of a water E supply or of a public sewer from the Water Authority under the Water Acts 1945 and 1973, or agreements made under the Public Health Act 1936.
>
> . . .
>
> **General policy**
>
> 4. It is a matter of law as well as of good administration that planning applications should be considered on their merits having regard to the provisions of the development plan and any other material consideration and F that they should be refused only when this serves a clear planning purpose and the economic effects have been taken into account. (More detailed advice is given in DoE Circular 22/80 (WO Circular 40/80)). By the same token, the question of imposing a condition or obligation — whether negative or positive in character — should only arise where it is considered that it would not be reasonable to grant a permission in the terms sought which is not subject to such condition or obligation. A wholly unacceptable development G should not of course be permitted just because of extraneous benefits offered by the developer.
>
> 5. If a planning application is considered in this light it may be reasonable, depending on the circumstances, either to impose conditions on the grant of planning permission, or (where the authority's purpose cannot be achieved by means of a condition) to seek an agreement with the developer which would be associated with any permission granted. Such agreements may well assist towards securing the best use of land and a properly planned H environment. But this does not mean that an authority is entitled to treat an applicant's need for permission as an opportunity to obtain some extraneous benefit or advantage or as an opportunity to exact a payment for the benefit of ratepayers at large. Nor should the preparation of such an agreement be permitted to delay unduly the decision on the application.
>
> 6. The test of the reasonableness of imposing such obligations on developers depends substantially on whether what is required:

> A (1) is needed to enable the development to go ahead, eg provision of adequate access, water supply and sewerage and sewage disposal facilities (advice on the provision of infrastructure is given in Annex A to DoE Circular 22/80 (WO Circular 40/80) and on land drainage in DoE Circular 17/82 (WO Circular 15/82)); or
>
> (2) in the case of financial payments, will contribute to meeting the cost of providing such facilities in the near future; or
>
> B (3) is otherwise so directly related to the proposed development and to the use of land after its completion, that the development ought not to be permitted without it, eg the provision, whether by the developer or by the authority at the developer's expense, of car-parking in or near the development or of reasonable amounts of open space related to the development; or
>
> (4) is designed in the case of mixed development to secure an C acceptable balance of uses.
>
> Appendix B illustrates the application of these general principles to the provision of car parking.
>
> 7. If what is required or sought passes one of the tests set out in the preceding paragraph, a further test has to be applied. This is whether the *extent* of what is required or sought is fairly and reasonably related in case and kind to the proposed development. Thus, while the developer may D reasonably be expected to pay for or contribute to the cost of infrastructure which would not have been necessary but for his development, and while some public benefit may eventually accrue from this, his payments should be directly related in scale and kind to the benefit which the proposed development will derive from the facilities to be provided.
>
> 8. There is also a final test, namely whether what the developer is being asked to provide or help to finance represents in itself a reasonable charge on E the developer as distinct from being financed by national or local taxation or other means — eg as a charge on those using the facility provided. The essential principle to apply is that the facility to be provided or financed should be directly related to the development in question or the use of the land after development. It would not normally be reasonable, for example, to seek a contribution to road construction or improvement in the immediate vicinity of the proposed development unless the need for this arises wholly and substantially from the new development.

F The inspector took the circular as his guide and his conclusions in paras 20 and 22 are the result of applying the criteria laid down in the circular to this case. A reference by the inspector in para 20 to the funding of the bypass being more appropriately found from national and local taxation to which Mr Newberry took exception stems from para 8 of the circular.

Mr Newberry has failed to convince me that the inclusion of the bypass in G the application for permission makes a material difference to the way in which planning gain should be approached. An application which includes two elements, one of which is claimed as providing a benefit which can outweigh planning objections to the other element necessarily invites consideration as to whether the two elements are directly related from a planning point of view. Here the inspector has made a finding, which is one of fact, that the bypass is extraneous to the residential development. The two elements are not directly H related and any benefit from the one element is, therefore, not relevant to the issue of planning permission for the other. Mr Newberry says the inspector should have struck a balance between the two elements of the application, but, in my view, that does not arise if the two elements are not directly related.

I think also that there is force in Mr Keene's point that there is no planning gain in the mere grant of planning permission for the bypass. Any benefit lies in advancing the timetable for the bypass and paying for it. Those benefits, if they arise, are the result not of the inclusion of the bypass in the application but A of the collateral agreements with the highway authority.

Mr Newberry challenged the inspector's findings in paras 22, 23 and 24 on the timing of the bypass. He said the inspector had no evidence on which to base his conclusion that the county council would build the bypass themselves just as quickly in any event, but I note that the inspector had evidence from a Mr Link, a senior project engineer with the county surveyor's department of West Sussex County Council who attended at the inspector's request. So there B was evidence and I therefore reject this point of Mr Newberry.

At the end of his submissions in reply Mr Newberry, as something of an afterthought, said that the inspector should have considered granting planning permission for the bypass alone. I heard this point *de bene esse* as it is not in the notice of motion. I think it may well be that if two elements of a development in a single application for planning permission are not directly C related to one another, it is open to the local planning authority or the Secretary of State to treat them as the subject of two applications. This possibility was envisaged by Kerr LJ in the *Covent Garden* case [1990] 1 QB 87 at p 117C. But, in my view, it does not arise for consideration here, for the simple reason that the inspector was not asked to do it and it is too late to raise it now. In my judgment, therefore, there are no grounds for quashing this decision and the application must be dismissed. D

Applications for costs were made on behalf of the first and second respondents. Counsel for the local planning authority did not ask for a second set of costs but invited the deputy judge, in effect, to split one set of costs between the two respondents.

MR DAVID WIDDICOMBE QC: The application will be dismissed with one set of costs. As regards the split, I did get assistance from Mr Keene which is E reflected in the judgment, but I think what I have to look for is something special in the circumstances which justifies the award of costs to the planning authority as well as to the Secretary of State. It is a very common situation for there to be two respondents in these planning cases and I would want to find something, in the circumstances, which was special before I deprived the Secretary of State of costs and I do not find that here, so I give the Secretary of F State his costs of this application.

Application dismissed with costs.

R v Poole Borough Council, *ex parte* Beebee and others G

QUEEN'S BENCH DIVISION
Schiemann J
December 21 1990

Planning permission — Reasons for decision — Whether applicants had *locus standi* — Whether proper consideration given to SSSI status — Whether H consideration given to need for environmental assessment — Application dismissed

By a decision dated April 6 1989 the respondent council granted to themselves planning permission for housing development on part of Canford Heath, Dorset. In November 1988 the Nature Conservancy Council ("NCC") had notified the subject land to include it within a site of special scientific interest

A ("SSSI") to provide protection for, *inter alia,* smooth snakes, sand lizards, the Dartford warbler, the nightjar and the hobby. There had been an earlier outline planning permission for residential development which had expired.

The applicants, representing the Worldwide Fund for Nature (UK) ("WWF") and the British Herpetological Society ("BHS"), moved for judicial review of the council's decision on the grounds, *inter alia,* that they had failed to consider the fact that the subject site was part of an SSSI;
B improperly failed to consider whether an environmental impact assessment should be carried out prior to the grant of planning permission; and given undue presumption in favour of the grant of planning permission by reason of the 1984 permission. It was submitted on behalf of the respondent council that the applicants had insufficient interest to make the application.

Held The application was dismissed.

Although those representing WWF would not alone have had sufficient
C interest, the BHS had by reason of its financial input into the site and its connection with the planning permission and of a condition therein relating to the protection of species: *R* v *Secretary of State for the Environment, ex parte Rose Theatre Trust Co* [1990] 1 PLR 39 distinguished: see p 30B. Parliament has not imposed a duty on local planning authorities to give reasons for the grant of planning permissions: see p 30H. Although the council may have proceeded on the basis that the NCC merely proposed SSSI status, whereas the
D site had that status, the officer preparing the report and the council did take into account the substance of what they were required to take into account. The fact that the council may not have been aware that the site had been formally notified did not deprive them of knowledge of any material underlying fact: see pp 33-34.

The purpose of an environmental impact statement under the Town and Country Planning (Assessment of Environmental Effects) Regulations 1988 is to draw to the attention of a local planning authority material relevant to
E coming to a decision. In the present case the appropriate bodies, the NCC and the BHS, drew the significant factors to the attention of the council so that the council had in their possession the substance of what they would have had if they had considered the 1988 regulations and prepared a statement: see p 35H. The respondents had not given excessive weight to the 1984 permission. It was part of the planning background and the respondents had grappled at some length with the planning/historical arguments in favour of permission and the
F strong conservation arguments against. They had not taken into account any irrelevant matter: see pp 36-37.

Cases referred to in the judgment

R v *Inland Revenue Commissioners, ex parte National Federation of Self-Employed and Small Businesses Ltd* [1982] AC 617; [1981] 2 WLR 722; *sub nom Inland Revenue Commissioners* v *National Federation of Self-Employed and Small Businesses Ltd* [1981] 2 All ER 93, HL
G *R* v *Secretary of State for the Environment, ex parte Rose Theatre Trust Co* [1990] 1 QB 504; [1990] 2 WLR 186; [1990] 1 All ER 754; [1990] 1 PLR 39

Application for judicial review

This was an application for judicial review of a decision by the respondents, Poole Borough Council, to grant to themselves planning permission.

Gerard Ryan QC (instructed by Cyril Clark & Son, of Bournemouth) appeared for
H *the applicants, Trevor Beebee, William Whitaker, George Medley and Michael Dopson.*

Jeremy Sullivan QC and Peter Village (instructed by the solicitor to Poole Borough Council) appeared for the respondents.

The following judgment was delivered.

SCHIEMANN J: This is an application for judicial review of what in substance is a grant on April 6 1989 by Poole Borough Council to themselves of planning permission for housing development of some land which is part of A Canford Heath and to which I shall refer as the subject sites. The subject sites have a long history of being seen as appropriate for development of this kind. They used not to be part of a site of special scientific interest ("SSSI") under the terms of the Wildlife and Countryside Act of 1981, although much of Canford Heath was such a site.

However, on November 7 1988 the Nature Conservancy Council ("NCC") notified the council that they had revised the boundaries of the SSSI so as to B include the subject sites. The main reason for the revision of the boundaries was that the sites support a number of protected species, including smooth snakes, sand lizards, the Dartford warbler, the nightjar and the hobby.

The effect of this notification in broad terms was to oblige the council to consult the NCC before doing anything in respect of the subject sites which might harm the protected species. This the council did. The NCC objected to C the housing proposal and asked the Secretary of State for the Environment to call it in for his decision. The Secretary of State refused to do so, and so the council made the planning decision themselves and in effect granted themselves planning permission.

Standing

The applicants tried to persuade the NCC to apply for leave to move for D judicial review of the council's decision but the NCC was not prepared to act. So the applicants applied instead.

Pursuant to section 31(3) of the Supreme Court Act 1981, I have considered whether the applicants have a sufficient interest in the matter to which the application relates. The applicants are two pairs of individuals, one pair claiming to represent the Worldwide Fund for Nature (UK) and the other the British Herpetological Society ("BHS"), both unincorporated associations. E

The BHS has a long-established association with the subject sites; much of its work in the field has been funded by the NCC and it has carried out work at Canford Heath. It has had a continuing and genuine interest in the subject sites for years, which has been recognised by the council. Indeed the sixth condition of the planning permission which the council granted themselves provides "that prior to any development starting on the site a full season's notice shall be given to the BHS to enable the catching and relocation of rare F species known to inhabit the site". This is a condition said to be imposed in the interests of conservation and to conform with the Wildlife and Countryside Act. The position of the Worldwide Fund for Nature (the current name for the World Wildlife Fund) is that it had been involved in the conservation of the Dorset heathlands for over 15 years, has made grants to the BHS since 1971 to assist them in carrying out their work on habitats for sand lizards and smooth G snakes, and is an accredited participant at meetings of the contracting parties to the Berne Convention since 1986.

Mr Sullivan QC, who appears on behalf of the respondents, submits that neither of these associations has a sufficient interest. He relies on the fact that they have no legal interest in the land, or in any neighbouring land, and on the fact that the NCC has been given a role by Parliament to play in relation to SSSIs. He submits that where, as here, Parliament has set up a body such as the H NCC specifically to protect the interest which is sought to be protected by the application, the court ought to be slow to find that others have a sufficient interest. I have some sympathy with this submission.

Further, he relies on the approach which I adopted in *R* v *Secretary of State for the Environment, ex parte Rose Theatre Trust Co* [1990] 1 QB 504* at

*Editor's note: Also reported at [1990] 1 PLR 39.

A p 518. My decision in that case has not escaped academic criticism (see for instance [1990] *Public Law* 307 and [1990] *Cambridge Law Journal* 189) and the question of *locus standi* is something of a legal minefield following the five separate judgments on the point delivered in the House of Lords in *R* v *Inland Revenue Commissioners, ex parte National Federation of Self-Employed and Small Businesses Ltd* [1982] AC 617.

The Worldwide Fund for Nature made no representations in relation to the
B grant of planning permission and indeed were not a party to the original application for leave to move for judicial review. They were joined by consent on July 16 1990 upon giving an undertaking to pay the respondents' costs in the event of the applicants being ordered to do so. It seems to me that the BHS, with its long association with this site, its financial input into the site and its connection with the planning permission by being named therein, has
C sufficient interest for me to be entitled to look at its application on its merits. This I now do. Had the Worldwide Fund for Nature stood on their own, or had there been no consent order, I would not have considered them to have had a sufficient interest, but in the circumstances I do.

Grounds of challenge

I turn now to the grounds of challenge. The applicants' grounds of challenge
D have in the course of the history of these proceedings expanded and contracted from time to time, but at the conclusion of the proceedings they were in substance threefold.

1. The council did not take into account a relevant consideration, namely the fact that the subject sites were part of an SSSI.

2. The council improperly failed to consider whether an environmental impact assessment should be carried out in relation to the subject sites prior to
E the grant of planning permission.

3. The council gave excessive weight to the fact that in 1984 residential planning permission had been granted for the subject site and wrongly applied an alleged presumption in favour of the grant of planning permission.

The statute obliges local planning authorities to give reasons for *refusing* planning permission but does not oblige them to give reasons for *granting* planning permission. Statute gives those who have been refused planning
F permission a right of appeal on the merits but gives no right of appeal to those who disapprove of the grant of planning permission to another. Whereas in some countries there is a legal obligation on any public decision-making authority to set out in any decisions their motivation — that is the matters which they have taken into account — there is no such general rule in this country. For most of the history of the common law the courts have been reluctant to allow challenges to decisions which seem all right on their face.
G Parliament has stepped in on a piecemeal basis very often so as to enact that authorities give reasons for this or that decision. Such decisions are then easily challengeable on the basis that they betray an error of law on their face.

It is noticeable that in the case of local planning authorities who grant planning permission no such duty has been imposed by Parliament. In practice local planning authorities do not give reasons for the grant of
H planning permission and the relevant resolutions are, invariably in my experience, framed in an unmotivated form. Nor is there in town and country planning legislation, what can be found in some other statutes, a provision whereby any person affected by an unmotivated decision can write and ask for reasons.

Such unmotivated decisions are extremely difficult to attack on the basis that the decision-maker wrongly took into account matters which he ought not to have taken into account or vice versa. It is easy to be suspicious but it is

difficult to demonstrate that suspicion to be well founded. This is a well- A
known problem which, as I have said, has repeatedly been addressed by Parliament. Where reasons do not have to be given, as for instance in jury verdicts, it seems to be the consensus that society would rather live with the possible injustices of not knowing the reasoning of the decision-making body than with the inconvenience of forcing them to articulate reasons which may very well differ from one member of that corporate body to another.

Mr Ryan QC, who appeared for the applicants, was faced with an B
unmotivated decision, unexceptionable on its face. He did not allege bad faith or corruption or that the decision was on its face so wildly surprising that one's instinctive reaction would be that something must have gone wrong. No. He took me instead on a trawl of a considerable amount of predecision documentation, from letters written to an objector by an officer in the planning department via the report prepared by an assistant planning officer, C
upon which that officer elaborated in committee to various affidavits setting out who said what in committee. There is undoubted precedent for carrying out this sort of exercise without protest by the court, and indeed at times the courts have criticised authorities for not filing affidavits. So I criticise no one for having embarked on this exercise. If it is going to be done at all, it would be difficult for it to be done more expeditiously or elegantly than Mr Ryan did it.

However, for my part, I have grave reservations about the usefulness of this D
sort of exercise when there is no allegation of bad faith. These reservations, in part, arise out of the theoretical difficulties of establishing the reasoning process of a corporate body which acts by resolution. All one knows is that at the second that the resolution was passed the majority were prepared to vote for it. Even in the case of an individual who expressly gave his reasons in council half an hour before, he may well have changed them because of what was said subsequently in debate. It is that type of consideration, coupled with E
the fact that many of those who vote on a resolution may give no utterance in debate, which has led our courts not to permit references to Parliamentary debates when arguing about the meaning of a statute.

In part also, my reservations arise from the very fact that Parliament has not imposed a duty to give reasons on local planning authorities when granting planning permission, although it is well known that the absence of recorded reasons poses difficulties for anyone who wishes to challenge decisions on the F
basis of defective ratiocination. The problem cannot have been overlooked in the context of planning, because express provision is made for the giving of reasons if permissions are to be refused. All this may well point to a desire on Parliament's part not to have the implementation of positive decisions held up by legal challenge on the basis of defective reasoning. Be that as it may, I record the most significant matter which emerged from this exercise upon G
which we engaged for a couple of days.

There was before the relevant committee an officer's report prepared by Mr Garrard, the council's assistant planning officer, which describes the site and contains the following passages:

> The current proposal is to seek outline approval for residential development.
>
> The site history includes an outline approval for land including the subject H
> sites which was issued in 1984.
>
> The subject sites were included in the neighbourhood plan for Canford Heath with an SSSI to the north. The NCC have however declared an intention to extend the boundary of the SSSI which encompasses North Canford Heath to include the subject sites. This was notified to the Council on 7th November 88. In a response dated 1st March 1989 the Council registered an objection to this extension onto Canford Heath.

A Representations were set out thereafter, including ones by, first, the RSPB, which lodged an objection on the basis of the whole of Canford Heath having been made a candidate for a special protection area under an EC directive and pointed out that potential special protection areas ("SPA") under the directive should be treated as declared SPAs. Mr Garrard pointed out that the advice of the directive is that planning permission should be granted only when the authority are satisfied that disturbance or change to habitats will not be
B significant. Since the area is the nesting location of at least two birds in the directive's appendix, the RSPB lodged an objection and urged the council to conserve these areas for the import of wild life they hold.

Second, the NCC lodged an objection stating that the areas were of great conservation interest with a wide range of heathland types and varied topographies and that the "range of plants now supported here is thus
C exceptionally rich". This is further supported by a second letter stressing the importance of the habitat and the range of species on the land. It considered that if the council were minded to approve these developments then the NCC would seek to ask the Secretary of State to call in the application.

Mr Garrard pointed out that the BHS had lodged a general objection and that the county planning officer had stated that the development would conflict with policy 7.11 of the structure plan, as it now lay within the
D proposed SSSI boundary, and therefore further consideration should be given to the needs of this development.

There then follows a page headed "Planning Considerations", which reads:

> This land has always been included within the area intended for development within the Neighbourhood Plan for Canford Heath. This has been confirmed as recently as 1984 when an outline permission was granted
> E for the whole of Central Canford Heath, including this site (Reference 26567/2). There is then a full presumption in favour of renewing an outline unless there are overriding reasons for not so doing. Here the only change is that of the NCC proposing to extend the SSSI into the known development area in the interest of nature conservancy and being supported by a number of other Conservation Bodies.
>
> The wholesale objection of these Bodies to this application is unreasonable
> F in one sense in that the land has been held for residential development for some long time and the original boundaries of the SSSI and North Canford Heath were drawn with this knowledge. To extend the SSSI at this late date seems unreasonable, particularly as indicated to the NCC in the letter of the 1 March 1989 which stated:
>
> "The NCC were well aware of the residential development planned for this area in 1984 and circumstances have not altered since then."
>
> G Against that is the growing appreciation of the ecological value, and the scarcity of heathland, and the recognition of this in national and international legislation.
>
> However, in an attempt to accommodate the needs of the Conservation Bodies, the applicant is prepared to dedicate a large area of open heathland stretching southwards out of North Canford Heath on a high south facing ridge line. This is the most ecologically valuable area of the site and is a valuable contribution, since it is an area which has hitherto formed part of the
> H development area and enjoys the benefit of the 1984 permission.
>
> The remaining land is less ecologically valuable, but should be subject to careful examinations and rescue of protected species prior to any development.
>
> In the light of the long held presumption in favour of development and in the light that the circumstances have not altered since the NCC declared the original boundaries of the SSSI, the application should be agreed in outline, subject to the proposed tongue of heathland being retained, clearance of the

> rest of the land in association with the BHS, and the position of an adequate A area of open space within the residential area.

Mr Garrard has sworn several affidavits. He states that on the morning of the meeting the town clerk's representative informed the subcommittee that a telephone message had been received from the Department of the Environment requesting that the applications be not determined until the department had considered whether to call them in. A further message was B received later during the day indicating that the department did not propose to intervene. He points out that the agenda paper for meetings of the committee is sent to all members of the council whether or not they are members of the committee. He points out that the letters of representation were available and he says that he took particular care, when orally presenting this particular item to the committee, to refer to the representations in some detail. He said he referred to the reasons for the notification of Canford Heath as an SSSI and to C its international significance, that he noted its status as a proposed special protection area for birds and that he emphasised the increasing importance of ecological issues. He says the debate lasted some half an hour or so.

A solicitor of the council who was present in effect confirms this and states that Mr Garrard commented that the council had made investment decisions and should generally be entitled to rely on the development plans for certainty in those decisions. He says that in summarising his presentation Mr Garrard D said that there was a balance to be struck. There was a strong policy/historical argument in favour of grant and strong conservation reasons against the grant. On balance he recommended that the application be approved. He says the whole thrust of the debate in council was centred on the ecological issues on the one hand and the housing need and policy issues on the other. Eventually there was a unanimous decision for granting permission. E

I turn now to examine the three grounds of challenge, which I have identified earlier in this judgment. They were not argued in this order but to treat them thus will I hope make this judgment easier to understand.

First, the council did not take into account a relevant consideration, namely the fact the subject sites were part of an SSSI. That is the first submission. It is common ground that the subject sites were part of an SSSI and that this was a relevant consideration. Further, it is common ground that the council knew F that the NCC wished to extend the boundary of the existing SSSI so as to include the subject sites and apparently took that fact into account.

The wording of the officer's report with its references to "intention to extend the boundary of the SSSI" and "proposed SSSI boundary" is unfortunate and supports the applicants' contention that the council may well have proceeded on the basis that the subject sites were not as yet within an SSSI. Hence Mr Ryan's submission that the council failed to take into G consideration a relevant consideration, namely that the subject sites were within an SSSI. I regard that submission as justified.

However, although in general, once the submission has been accepted that the decision-maker failed to take into account a relevant consideration, the decision will be quashed, I do not think that that result should follow in the instant case. My reason for so holding is that, in my judgment, the officer H preparing the report, and the council, did take into account the substance of what they were required to take into account.

The procedure for establishing an SSSI under section 28 of the Wildlife and Countryside Act 1981 used to be that the NCC gave notice of a proposed notification to the owner, invited him to make representations, considered his representations and then decided whether to notify the owner that it considered the site to be one of special scientific interest and specifying

A operations which appeared to the NCC to be likely to damage the flora or fauna on that site. If the owner, having been notified of the NCC's decision, thereafter carried out any notified operations without consent, he committed an offence. Further, the local planning authority were under an obligation to consult the NCC prior to granting any planning permission in relation to such a notified site. That regime proved to be defective because some owners who had been served with a notice of proposed notification thereupon set to and B destroyed the site without waiting to see whether the notification was ever made.

By the Wildlife and Countryside (Amendment) Act of 1985 this loophole was closed and a different regime was established. Nowadays the site is protected while the NCC is considering representations. That protection is withdrawn if the NCC is persuaded by objections that the site should not be protected. While the site is protected the local planning authority are obliged C to consult the NCC prior to granting any planning permission. The purpose of the old regime and the new regime, so far as presently relevant, is to ensure that the planning authority, prior to granting permission, are apprised of the fact that the NCC regards the site as one of special scientific interest. In the present case there is no doubt that the planning authority were perfectly well aware of this fact and were aware of the reasons for it. The fact that the planning D authority may not have been aware that the site had been formally notified under the new regime did not deprive them of knowledge of any material underlying fact.

Although Mr Ryan correctly points out that the officer's report at one point states that "circumstances have not altered since the NCC declared the original boundaries of the SSSI", in my judgment the committee manifestly appreciated that which in fact had happened, namely that the NCC had E formed the view that its original boundaries for the SSSI were inadequate and should be extended in such a way as to encompass the subject sites. There is, in my judgment, nothing improper in the officer observing in his report, "to extend the SSSI at this late date seems unreasonable", provided that he balances it, as he does, with the assertion that "against that is the growing appreciation of the ecological value, and the scarcity of heathland, and the recognition of this in national and international legislation".

F I turn to the second submission. This was that the council improperly failed to consider whether an environmental impact assessment should be carried out in relation to the subject sites prior to the grant of planning permission. It is common ground that this was not considered by the council, and it is common ground that they were not obliged as a matter of law to carry out such an environmental impact assessment. However, Mr Ryan submits that they should have considered whether to carry out such an environmental impact G assessment and that their failure so to consider should lead to the permissions being quashed.

I started this judgment by referring to what I described as in substance being a grant by the council to themselves of planning permission. In order to explain the submission with which I am now concerned, it is necessary to be a bit more technical. The Town and Country Planning General Regulations 1976 (SI 1976 No 1419) make provisions for circumstances in which the H planning authority seek to obtain permission for development of land in their area which is vested in them but which development they do not themselves propose to carry out. In such circumstances, which applied in the present case, regulation 5(2) provides that the authority shall pass a resolution to seek permission for the development. (This resolution is commonly referred to as the first resolution.) There are then various provisions for advertisements and notifying a variety of people so that they may object. There is then a provision

contained in regulation 4(4), as applied by regulation 5(4), for a second A
resolution to resolve, which authorises the carrying out of the development. There is an opportunity for the Secretary of State to call the matter in to be dealt with by himself where he deems appropriate. That did not happen in this case.

The submission with which I am now concerned arises under the Town and Country Planning (Assessment of Environmental Effects) Regulations 1988
(SI 1988 No 1199). Regulation 17 provides in para (1): B

> Where this paragraph applies an authority shall not —
> (i) pass a (first resolution) . . . without first preparing an environmental statement in respect of the proposed development; or
> (ii) pass a (second resolution) . . . without first taking the environmental information into consideration.

Subpara (ii) indicates that subpara (i) applies where an application for C
planning permission for the relevant development would be a "Schedule 2 application".

"Schedule 2 application" is a term of art and refers to "an application for planning permission (other than an application made pursuant to section 31A) for the carrying out of development of any description mentioned in Schedule 2 which would be likely to have significant effects on the environment by
virtue of factors such as its nature, size or location". Schedule 2 to those D
regulations indicates a whole series of developments which have the potential of creating environmental problems. The only ones conceivably relevant are those contained in para 10, headed *"Infrastructure projects"*, and in particular subpara (b) "an urban development project".

The "environmental statement" referred to in regulation 17(1)(i) is defined as meaning such statement as is described in Schedule 3 to those regulations.
That schedule says that an environmental statement comprises a document or E
series of documents providing, for the purpose of assessing the likely impact upon the environment of the development proposed to be carried out, the information specified in para 2. That information involves a description of the development proposed, a description of the likely significant effects direct and indirect on the environment of the development, explained by reference to its possible impact on human beings, flora, fauna, soil, water, air, climate,
landscape, and so forth. F

The "environmental information" referred to in regulation 17(1)(ii) is defined as meaning the environmental statement and any representations in relation to the proposed development.

Although shortly after the first resolution was passed the BHS was informed, it never then took the point that an environmental statement was required. Mr Ryan concedes this but says that it is irrelevant because the
authority should consider these things whether or not it is drawn to their G
attention. He does not, however, seek to quash the first resolution on this basis.

Mr Sullivan submits that this was not a case where an environmental assessment was required because, he submits, this was not a case of an urban development project, and in any event the development was unlikely to have
significant effects on the environment. It is conceded by Mr Ryan that this is a H
view which is open to the authority but, he says, the authority never considered the regulations at all. In this he in turn is conceded to be right.

In my judgment, this point can be disposed of when one remembers that the purpose, in circumstances such as the present, of any environmental assessment is to draw to the attention of the authority material relevant to the coming to a decision. In the present case, it seems to me that the appropriate bodies, the NCC and the BHS, drew the significant factors to the attention of

A the authority, and so the authority had in their possession the substance of what they would have had if they had applied their minds to the 1988 regulations and had prepared such an environmental statement. The substance of all the environmental information which was likely to emerge from going through the formal process envisaged by the regulations had already emerged and was apparently present in the council's mind. In those circumstances it would be wrong to quash either of their resolutions on the
B basis of this submission.

I turn to the third one. This is that the council gave excessive weight to the fact that in 1984 residential planning permission had been granted for the subject sites and wrongly applied an alleged presumption in favour of the grant of planning permission. Mr Ryan submits that this case raises a point of general concern as to the proper approach to a planning application in an area
C which has the highest status in conservation terms. He points out that the site here is (a) an SSSI under the Wildlife and Countryside Act 1981; (b) part of a candidate special protection area under section 3 of the 1981 Act and article 4 of the EC Birds Directive; and (c) the home of several species protected under Part 1 of the 1981 Act, the Birds Directive and the Berne Convention.

He submits that, given those statutory and policy considerations, it was an error for the council to treat this as a site where any presumption in favour of
D planning permission applied.

In substance, he submits that this application should have been considered on its merits and he submits that instead it laboured under a substantial disadvantage in having a presumption erected from the previous planning history. He does not, and could not, contend that the previous planning history is irrelevant. The background to his submission is that on April 25 1984 permission had been granted to develop a large part of Canford Heath
E including the subject sites for residential purposes.

That permission was granted subject to two conditions of present relevance:

> 2. Application for approval of any "Reserved Matter" must be made not later than the expiration of three years beginning with the date of this permission.
>
> 3. The development to which this permission relates must be begun not later than whichever is the later of the following dates:
>
F
> (i) the expiration of five years from the date of grant of outline planning permission or,
>
> (ii) the expiration of two years from the final approval of the Reserved Matters . . .

Mr Ryan makes the point that at the time of the consideration of this matter by the development control subcommittee on April 6 1989, the outline
G permission to which I have referred could not be implemented without further ado because the three-year time-limit for the application for approval of the reserved matter had elapsed. He concedes that, by virtue of section 31(A) of the Town and Country Planning Act 1971, application could have been made for the development of the land without complying with this condition but points out that it never was made, still less granted. In those circumstances he submits that the officer advising the committee erred in asserting that the
H subject sites enjoyed the benefit of the 1984 permission. The point is of importance, he submits, because the officer's report, after referring to the 1984 outline permission granted for the whole of central Canford Heath, including this site, continues:

> There is then a full presumption in favour of renewing an outline (sc. permission) unless there are overriding reasons for not so doing. Here the only change is that of the NCC proposing to extend the SSSI into the known development area in the interests of nature conservancy.

Later in the report he refers to the subject sites as enjoying the benefit of the 1984 permission and also refers to the long-held presumption in favour of development. Mr Ryan submits that the committee were misled as to the status in law of the 1984 permission.

It is worth reminding oneself that the committee were not concerned with any attempt to implement the 1984 permission but rather with an application for a new permission. The 1984 permission was relevant as part of the planning background, as was the fact that in the planning of the area it had long been assumed that these sites would go for residential development. The essence of Mr Ryan's complaint does not I think turn on the technicality as to whether the 1984 permission was at the date of the report still capable of being implemented but rather on what he says is a misleading approach to be found in the report and its references to presumption.

Having looked at all the evidence, I am not persuaded that the committee here were misled at all. Planning circulars and structure plans are full of broad statements with many presumptions, many of which are mutually irreconcilable, so that in a particular case one must give way to another. Neither the instant report nor the planning circulars are to be read as though they were taxation statutes to be carefully construed. The local authority officers who draft these reports are not by training or inclination in general endowed with the skill of Parliamentary draftsmen. It seems to me that Mr Garrard was absolutely right when he said, as appears from the council's solicitor's affidavit, that there was a balance to be struck. There were strong policy/historical arguments in favour of a grant and strong conservation reasons against the grant. That was the matter with which the committee grappled at some length and came down in favour of adhering to their earlier policy, notwithstanding the fact that they knew that the NCC were anxious that this site should be preserved for perfectly good ecological reasons.

In my judgment, when one looks at the evidence as a whole rather than the odd isolated phrase in one or another of the documents before the council, there is no reason to suppose that the council took into account any significant matter which they should not have taken into account; or that they failed to take into account any significant matter which they ought to have taken into account; or that they committed some other error which should lead to this permission being quashed. Some of the criticisms made by Mr Ryan of this phrase or that do have force. But at the end of the day the court has to sit back, as it were, and ask itself whether there was anything which vitiated the decision-making process and which would make it desirable for the council to look at this matter again.

In my judgment, there was not, and this application is dismissed.

Application dismissed with costs.

Cotterell and others *v* Secretary of State for the Environment and another

QUEEN'S BENCH DIVISION

Mr Roy Vandermeer QC (sitting as a deputy judge of the Queen's Bench Division)

January 14 1991

Bias — Inspector drinking in public house with representatives of parties after site inspection — Planning officer and objector staying for further drink with

A **inspector after appellants' representatives left — Whether decision to be set aside — Whether reasonable impression of bias**

The applicants appealed against the refusal of their applications for planning
permission for residential development on five separate sites in Hereford. A
public inquiry was held into the appeals and the following day the inspector
carried out his site inspections. At four of the sites he was accompanied by Sir
B John Cotterell, representing the appellants, by the appellants' planning
consultant, Mr Wood, and by the planning officer who had given evidence for
the local planning authority Mrs Holborow. At the fifth site at Staunton-on-
Wye they were joined by a local resident, Mr Fraser, whose property adjoined
the appeal site and who objected to the proposed development. Following the
final site inspection the inspector invited the four people who had
accompanied him to join him for a drink in a local public house. The inspector
C paid for the drinks. An offer by Sir John to buy everyone a second drink was
declined, but the inspector said that he would stay at the public house for
lunch. Sir John and Mr Wood then left, expecting that Mrs Holborow and Mr
Fraser would also leave soon afterwards. However, they stayed for about
another 20 minutes during which time the inspector bought a further round of
drinks. They did not talk about the inquiry but matters of common interest
were discussed, including the fact that the inspector and Mr Fraser had been to
the same college. The inspector dismissed all the appeals. The applicants
D sought to quash the decision on the ground of apparent bias.

Held The application was dismissed.
1. The test to be applied was whether what had happened was something
which could have led somebody to the not unreasonable conclusion that
something was being done to interfere with the natural course of justice: see
p 41E. In considering this question, regard should be had to the content of
E the conversation which took place in the absence of the appellants'
representatives as deposed to in the affidavits: see pp 41F-42C.
2. It was important to bear in mind that casual conversations did happen by
chance when someone on the other side was not there and this alone would not
amount to a procedural impropriety or give the appearance of it: see p 42G.
3. On balance, considering the whole of the background, the nature of the
conversation that had gone on before Sir John and Mr Wood left, and the fact
F that it was at least known to them that the planning officer and Mr Fraser
would stay on, albeit not for as long as they in fact did, the inspector's conduct
fell on the acceptable side of the line: see pp 43H-44A. The facts could be
distinguished from those in *Simmons* v *Secretary of State for the Environment*
[1985] JPL 253, *Furmston* v *Secretary of State for the Environment* [1983]
JPL 49 and *British Muslims Association* v *Secretary of State for the
Environment* (1987) 55 P&CR 205.

G **Cases referred to in the judgment**
British Muslims Association v *Secretary of State for the Environment* (1987) 55 P&CR 205
Furmston v *Secretary of State for the Environment* [1983] JPL 49
Metropolitan Properties Co (FGC) Ltd v *Lannon* [1969] 1 QB 577; [1968] 3 WLR 694; [1968] 3 All ER 304, CA
Simmons v *Secretary of State for the Environment* [1985] JPL 253
Steeples v *Derbyshire County Council* [1985] 1 WLR 256; [1984] 3 All ER 468; [1981]
H JPL 354

Application under section 288 of the Town and Country Planning Act 1990
This was an application under section 288 of the Town and Country Planning Act
1990 by the applicants, Henry Cotterell, Brian Stevens, Guy Smallwood and
Andrew Arkwright, to quash a decision of the first respondent, the Secretary of
State for the Environment, by his inspector, whereby he had dismissed five
appeals against decisions of the second respondents, Leominster District Council,

who had refused applications for planning permission for residential A
development.

Martin Kingston (instructed by T A Matthews & Co, of Hereford) appeared for the applicants, Henry Cotterell, Brian Stevens, Guy Smallwood and Andrew Arkwright.

Michael Kent (instructed by the Treasury Solicitor) appeared for the first respondent, the Secretary of State for the Environment.

The second respondents, Leominster District Council, did not appear and were B
not represented.

The following judgment was delivered.

MR ROY VANDERMEER QC: In this case the applicants apply under what is now section 288 of the Town and Country Planning Act 1990 to quash a decision of the Secretary of State for the Environment made on his behalf by one of his inspectors. The decision was given in a letter dated August 16 1990. C

The hearing and the decision letter related to a number of appeals, some four relating to land at Byford Common, Hereford, and the fifth relating to land adjacent to Hempridge, Staunton-on-Wye, Hereford. All five appeals were dismissed. Perhaps somewhat curiously, the facts of the case and indeed of the inspector's decision letter are not of any moment. No complaint is made about the contents of the decision letter itself. What is complained about is
what is said to be an apparent bias — I put it in that shorthand for the moment D
— on the part of the inspector.

Before expanding upon that, it is convenient if I recite the relevant facts. The inquiry, which was apparently to have taken two days, sat for what I must confess seems an astonishingly long time on the first day and was, so far as the hearing was concerned, disposed of on that day. There was, however, the need for a site inspection to take place in relation to each of the sites and that was E
arranged for the following day.

On the following day the various parties gathered and there was initially a site inspection of the four sites which were the subject of the first four appeals, all of which, as I have already indicated, were at Byford Common. Following the completion of the site visits with regard to those appeals, about which there is no complaint whatsoever, the inspector set off to Hempridge. Sir John
Cotterell, representing the appellants, and the appellants' planning F
consultant, Mr Richard Wood, also went to Hempridge, as did the planning officer who had given evidence for the local planning authority, Mrs Holborow. They travelled independently and arrived at the site where they met a Mr Murray Fraser, a local resident who occupied a dwelling next to the appeal site at Staunton-on-Wye and who objected to the development. There then took place a site inspection. The inspector looked at the site, studied the
village and walked around the area generally. No complaint is made of G
anything that occurred at that stage.

In due time, on what was apparently an extremely hot day, the site inspection finished and the inspector invited the four people who had accompanied him to join him in refreshment by way of a drink in a local public house. All four accepted the invitation and consequently the five people to
whom I have referred went to the public house and had a drink together. The H
inspector paid for the drinks, no doubt to maintain a position of independence.

During the course of the time when all five people were drinking together there was conversation. In one of the affidavits sworn by Sir John Cotterell there is an apparent suggestion that something was being discussed about the way that the inquiry itself had been conducted, albeit of a relatively innocent nature relating to the procedure. A similar indication appears from the other

A affidavits including, I think, Mr Wood's, but that matter is not raised as a point in this court. The conversation did not, in fact, touch upon the issues of the case, or indeed upon the case in any way, although there were some conversations relating to planning matters generally. But again no complaint is made about anything that occurred at the stage while all five persons were together.

Eventually Sir John Cotterell indicated that he had to leave. Certainly at
B about that time, either just before he indicated that he had to leave or possibly shortly thereafter, he offered to buy the other three people a drink — not Mr Wood, who was leaving with him, as I understand it. That offer was declined. Mr Wood, the appellants' planning witness, said, in para 5 of his affidavit, that it was after the party had been in the public house for about 15 minutes that Sir John asked everybody if they would like a second drink. He says that
C the offer was declined. He went on to say that it was then nearing one o'clock. Sir John said he needed to return to Garnons, which was the property he was looking after, to attend to estate matters there. Mr Wood then says that the inspector at that point mentioned that, rather than setting off for home and having lunch on the way, he would stay at the New Inn for lunch.

Mr Wood, in para 6 of his affidavit, said that his recollection was that Mrs Holborow and Mr Fraser had nearly finished their drinks. At the time that Sir
D John and Mr Wood decided to leave, the situation as to the other three was that the inspector had indicated that he intended or expected to stay in the public house for some time, while the planning officer and the local objector, Mr Fraser, were clearly not quite ready to go — they still having some drink, albeit a small amount, to consume. But, on the other hand, all three who remained had declined Sir John's offer of a drink and Mr Wood said that he anticipated that the planning officer and Mr Fraser would leave the New Inn at the same
E time as Sir John, although when he and Sir John actually left he appreciated (as he said at para 6) that they were not going to leave at the same time as Sir John and himself — they seemingly still had something to drink — but he thought they would follow immediately afterwards.

The situation on the facts, as I read them then, can be put in a fairly short way thus. By agreement, of which no complaint is made, these five people had
F a drink together. At the end of the time that they were together those representing the appellants left, knowing that the others were going to remain at least for a moment or two; not being clear as to precisely when the planning officer and Mr Fraser would be leaving but anticipating it would not be long after their own departure.

Up to that point Mr Kingston accepts that no complaint could be made and that complaint could not be made of the fact that the inspector, the planning
G officer and Mr Fraser remained at least for some short time together in the public house. In actual fact the inspector and the other two did not leave for some little time, possibly as long as about 20 minutes, and during that time, although a drink had been declined from Sir John, the inspector purchased further drinks for the others, no doubt still insisting that he took the lead so that others were not expending money on him. Mr Kingston says that that conduct should justify the decision being quashed.

H I turn briefly to the law. I can deal with it fairly briefly because it is essentially agreed between the parties. I was referred to a number of cases, the first of them being *Metropolitan Properties Co (FGC) Ltd* v *Lannon* [1968] 3 WLR 694. The circumstances of that case, let me say at once, were, in my judgment, clearly very different from the circumstances in this case. Nevertheless certain important principles were established.

Lord Denning MR, at p 707, made it clear that the court does not look to see if there was a real likelihood that the adjudicator — in that case a chairman of a tribunal — would or did in fact favour one side at the expense of the other, but he said: A

> The court looks at the impression which would be given to other people. Even if he was as impartial as could be, nevertheless if right-minded persons would think that, in the circumstances, there was a real likelihood of bias on his part, then he should not sit.

One can translate those words "he should not have acted in that way". In this case, on the basis of the judgment of the learned Master of the Rolls, and if right-minded people would think there was a real likelihood of bias by the contact that took place, then the decision cannot stand. B

However, Edmund Davies LJ (as he then was) examined what he referred to as the "real likelihood" test and, in effect, rejected it as establishing too high a standard before a decision should be quashed. He said at p 713C: C

> . . . I take the view that the requirement that justice must manifestly be done operates with undiminished force in cases where bias is alleged and that any development of law which appears to emasculate that requirement should be strongly resisted. That the different tests, even when applied to the same facts, may lead to different results is illustrated by *Reg* v *Barnsley Licensing Justices** itself, as Devlin LJ made clear in the passage I have quoted. But I cannot bring myself to hold that a decision may properly be allowed to stand even although there is *reasonable* suspicion of bias on the part of one or more members of the adjudicating body. D

Simmons v *Secretary of State for the Environment* [1985] JPL 253, the next case to which I was referred, brought that more clearly into the planning context and put the matter in a way that can, I think, be readily used as the yardstick for the judgment in this case. It does not in any way differ from what the Master of the Rolls and Edmund Davies LJ said in *Metropolitan Properties*, but in terms of the language, because it was a planning case, it comes extremely close to the situation here. E

I pose the test in this way. Looking at the facts of this case, "was what happened" (and I use Forbes J's words) "something which could have led somebody who knew the background to the not unreasonable conclusion that something was being done to interfere with the natural course of justice?" If that is the conclusion that I come to here then it follows that the decision must be quashed. As was said, I think it was, by Forbes J, but certainly accepted by both parties here, it is a matter of impression for me in this instance. F

Before I go on, against the background of that test, there is only one other matter of law which I need address. I have before me affidavits as to the nature of the conversation that occurred after the appellants' representatives had left. I do not intend to recite the whole detail of that, but what is clear is that certain matters of common interest and common ground were established between the inspector on the one hand and Mrs Holborow and Mr Fraser on the other. They did not relate to anything specific to the inquiries that had just been conducted but related to very general matters such as walking interests, and Mr Kingston particularly draws attention to the fact that the inspector had been to the same college for at least a period of time but at a very different period of time to Mr Fraser, and other mutual interests and points of contact of that sort. G H

Mr Kent, on behalf of the Secretary of State, says that I must ignore the knowledge of the conversation because I must look at it on the basis of the observer who saw and, I suppose, inferentially did not hear. On the other hand, Mr Kingston says that I ought to take account of that conversation and

*Editor's note: Reported at [1960] 2 QB 167.

A he draws my attention to a passage in the judgment of Webster J in the case of *Steeples* v *Derbyshire County Council* [1985] 1 WLR 256. The passage to which he referred me is at p 287C where Webster J said, referring to the hypothetical person referred to by Forbes J who knew the background, that he must be taken to know all matters, whether in fact known or available to the public or not, which are in evidence at the trial. Mr Kingston says, therefore, that I should take account of the contents of the affidavit and what was in fact B said between those three persons.

I have some hesitation in accepting that, but on balance I think it is right and as it is urged by the applicants in this case I propose, at least for the purpose of this judgment, to proceed upon the basis that that is the correct approach. It does, of course, cut two ways because what it demonstrates is that there was no discussion about the case. On the other hand it does show that there was some C mutuality of interest. Mr Kingston says it demonstrated a growing or potential sympathy between the inspector on the one hand and the objector and the planning officer on the other.

Taken together with the fact that they remained together, in the absence of the appellants' representatives, for some 20 minutes, shared a drink that was not expected by the appellants and their representatives to be shared and were establishing some kind of affinity, Mr Kingston urges that it would not be an D unreasonable conclusion for somebody knowing the background to think that there was at least the potential of bias, although of course it is accepted that there was in fact no bias or at least no conscious bias, Mr Kingston saying that bias can emerge quite unintentionally.

What is my impression of these matters? The starting point seems to me to be the acceptance that there was agreement for the parties to go together. It E started as a joint act and my view is that when Sir John and Mr Wood left it was known, albeit it was anticipated by them that it might be some short time, that the other parties would remain together. Thus, this is not a case where the inspector came together with the other parties in the absence of the appellants without at least their knowledge and potentially, possibly, even a measure of acceptance but acceptance, as Sir John and Mr Wood say, of a very limited nature.

F As I have said already, if the matter stopped there and had Mrs Holborow and Mr Fraser left very shortly after Sir John and Mr Wood, there would be no question of any offence to the principles of natural justice in this case and nor indeed would Mr Kingston, as he made plain to me, have made such a suggestion. Indeed, he went so far as to say, I think, that it would be perverse to make such a suggestion.

G I think in coming to a view it is also important to bear in mind the kind of comment made by Stuart-Smith J (as he then was) in the case of *British Muslims Association* v *Secretary of State for the Environment* (1987) 55 P&CR 205. In that case, at p 212, the learned judge accepted that there could be casual conversations which happened by chance when someone on the other side was not there, but in his view it could not possibly amount to a procedural impropriety or even give the appearance of it. He went on to point H out that one knows perfectly well that people move around on a site visit, as here, out of earshot because they are going to the lavatory or because the room is not big enough for them to be all there or something of that sort. In the circumstances of that case the learned judge came to the conclusion that that was not the situation that existed.

I do see, although it does not determine the matter, distinctions between a number of the cases put before me and the instant case. *Steeples*, of course, is a case of a quite different nature, as was the *Metropolitan Properties* case. However, *Simmons* is a case with some similarities. In that case, following the inquiry, the chairman of the planning committee, whom the appellant believed to be the principal instigator of the objection to the application, went to speak to the inspector with another person. While the learned judge concluded that there had been no impropriety on behalf of the inspector and that had it depended upon the question of whether the inspector was free from impropriety he would have had no hesitation in saying the appeal failed, he took the view that in the circumstances of that case it gave the wrong impression and accordingly that somebody who knew the background would take the view there was a reasonable possibility that there had been a breach of natural justice. A B

The primary distinction, in my view, as a matter of fact between that case and this is that in *Simmons* the inspector had been approached in the absence of the appellant and, as the judge found, allowed himself to talk for some little time with somebody whom the appellant believed to be the principal objector, albeit he had not actually, as I understand it, appeared at the inquiry. In those circumstances it would seem not unreasonable for this appellant to assume that there was some attempt by the chairman of the planning committee to press a point of view upon the inspector. C

The next case to which reference was made was a case referred to by Stuart-Smith J in the *British Muslims* case at p 211. That is the case of *Furmston* v *Secretary of State for the Environment* [1983] JPL 49. In that case, after the inquiry had finished — I think it was in fact after a site meeting — it was seen that the county council's representative remained talking to the inspector with an open file, which clearly could be taken to relate to the case. That conversation went on for some time. Again that is, in my view, not quite the same kind of facts as occurred in this case. In the *British Muslims* case again it is to be noted that the conversation that took place, and went on for some time, between representatives of the local planning authority and the inspector both started and was completed without the appellant being present, with some distance separating the two groups. Stuart-Smith J thought that it went on perhaps not as long as 20 minutes, whereas here it is suggested that the conversation went on for 20 minutes, and Stuart-Smith J did place some reliance on the fact that Mr Bajwa, the appellant, had some difficulty in language and that therefore the matter was of particular importance. D E F

There are distinctions in those cases, the fundamental one being that here the matter started with everybody being together and the inspector, the planning officer and Mr Fraser being left together with the knowledge and acceptance of the appellants' representatives, albeit they did not anticipate the inspector would remain with the others for so long. G

Which side of the line does it fall? I confess that at the outset of the case I felt some confidence that it fell on the right side of the line. I have listened with care to what Mr Kingston has said and I find some force in it. As I think he pointed out, had the inspector asked if it was all right if the three of them remained together and had another drink and received consent, the matter would have been beyond doubt. I do not find it wholly easy but I have come to the view that the inspector's conduct fell on the acceptable side of the line. In the circumstances of this case, bearing in mind the whole of the background, the nature of the conversation that had gone on before Mr Wood and Sir John left, the fact that it was at least known they would remain, albeit the amount of time they remained was to Sir John and Mr Wood surprising, I have come to the view with some hesitation that this is not a case whereby somebody who knew the background would reasonably reach the conclusion that there was H some apparent bias or that in the circumstances there was a breach of the rules of natural justice.

For those reasons this application is dismissed.

Application dismissed with costs.

Westminster City Council *v* Secretary of State for the Environment and another

QUEEN'S BENCH DIVISION

Mr Roy Vandermeer QC (sitting as a deputy judge of the Queen's Bench Division)

January 14 1991

Planning condition — Appeal by written representations — Inspector grants permission without condition — Effect of use of premises for temporary residential accommodation — Whether condition unnecessary — Whether section 25 of the Greater London Council (General Powers) Act 1973 provides control rendering condition unnecessary

On January 19 1989 the applicant council granted planning permission for the erection of a mews building for use as residential and professional offices. Permission was granted subject to three conditions. Condition 2 prohibited use of the premises (a) for temporary sleeping accommodation as defined in section 25 of the Greater London Council (General Powers) Act 1973 or (b) for holiday occupation. The second respondents' application to carry out the development without compliance with a further condition, condition 3, was refused. On appeal the inspector, after receiving additional representations, allowed the appeal discharging, *inter alia*, condition 2. The inspector concluded that the condition was unnecessary because of the powers contained in section 25 of the Greater London Council (General Powers) Act 1973*.

The council contended that condition (a) conferred an element of control over and above that conferred by section 25 of the 1973 Act and accordingly the inspector, who had relied on an earlier decision of the High Court, had erred.

Held The application was dismissed.

The condition did not provide an element of control in addition to section 25 of the 1973 Act. Section 25 applies where a building is erected with permission to be used as a residence and is first used as temporary accommodation. This is because the proper way to interpret the definition of "residential premises" in section 25(2)(*b*) is to ask whether the premises are capable of use as a permanent residence: see p 49. In the absence of some specific reference in the

*Editor's note: Section 25 of the Greater London Council (General Powers Act 1973, as amended, provides:

"(1) For the purposes of section 22(1) of the Act of 1971, the use as temporary sleeping accommodation of any residential premises in Greater London involves a material change of use of the premises and of each part thereof which is so used.

(2) In this section—

(*a*) "use as temporary sleeping accommodation" means use as sleeping accommodation which is occupied by the same person for less than ninety consecutive nights and which is provided (with or without other services) for a consideration arising either—

(i) by way of trade for money or money's worth; or

(ii) by reason of the employment of the occupant; whether or not the relationship of landlord and tenant is thereby created;

(*b*) "residential premises" means a building, or any part of a building, which was previously used, or was designed or constructed for use, as one or more permanent residences."

grant of planning permission to the right to use the premises for temporary A
purposes, it is unlawful so to use them even on a first use in the area to which section 25 applies: see p 50C. It would be a breach of planning control: see p 50D. If the building is first used for permanent residential accommodation and the use is changed to temporary residential accommodation, the use of the building remains as use of a dwellinghouse within Class C3 of the Use Classes Order 1987. The effect of section 25 of the 1973 Act is that what would otherwise have not required planning permission requires permission: see B
p 51B. It follows that the condition is unnecessary because its effect is produced by section 25: see p 51C.

Cases referred to in the judgment

Westminster City Council v *Secretary of State for the Environment* unreported, June 12 1989

Wilson v *West Sussex County Council* [1963] 2 QB 764; [1963] 2 WLR 669; [1963] 1 All ER 751; (1963) 61 LGR 287; 14 P&CR 301; [1963] EGD 565; 185 EG 683, CA C

Application under section 288 of the Town and Country Planning Act 1990

This was an application under section 288 of the Town and Country Planning Act 1990 by the applicant council to quash a decision of the first respondent, the Secretary of State for the Environment, by his inspector, who had allowed an appeal by the second respondents, Miller Developments Ltd, against a decision of the applicants not to allow development without compliance with a planning D
condition.

Michael Barnes QC (instructed by the solicitor to Westminster City Council) appeared for the applicants.

Guy Sankey (instructed by the Treasury Solicitor) appeared for the first respondent, the Secretary of State for the Environment.

The second respondents, Miller Developments Ltd, did not appear and were not represented. E

The following judgment was delivered.

MR ROY VANDERMEER QC: In this case the local planning authority, Westminster City Council, apply under what is now section 288 of the Town and Country Planning Act 1990 to quash a decision of the Secretary of State for the Environment given on his behalf by one of his inspectors. The decision that is challenged was given by letter dated April 17 1990. F

On January 19 1989 the local planning authority had granted planning permission for "the erection of a three-storey mews building for use as residential and car parking accommodation, use of main building as professional offices on basement, ground, first floor, second and third floors and residential flat on fourth floor, and installation of a lift pulley room and dome lantern light on the roof". Mr Barnes QC, on behalf of the applicants, G
pointed out that in fact planning permission was not required for the flat on the fourth floor of the existing premises, but nothing turns upon that matter and I simply record it as the factual position.

The grant of planning permission was subject to a number of conditions. On April 17 1989, before commencing the permitted development, the second respondents in this court, Miller Developments Ltd, applied to the council under what is now section 73 of the 1990 Act to carry out the development H
without complying with condition 3 of the planning permission.

That application was refused and an appeal was made to the Secretary of State under what is now section 78 of the 1990 Act. Although the appeal was only against condition 3, the effect of the appeal was to give to the Secretary of State the right to consider the other conditions subject to which consent had been granted. The appeal was conducted by way of written representations.

On February 20 1990 the planning inspectorate wrote to the parties pointing

A out that the inspector had noticed that condition 2 was virtually identical to a similar condition which he had discharged in an earlier appeal on the basis that it was unnecessary. It was pointed out that that decision had been unsuccessfully challenged in the High Court. The parties were accordingly invited to make representations with regard to condition 2 and that they did. The inspector allowed the appeal and granted planning permission in the form B originally granted by the local planning authority, save that he did not impose either the original condition 2 or the original condition 3. Nothing turns upon condition 3 and I shall therefore make no further reference to it.

Condition 2, which is to the point, had read as follows:

> The premises the subject of this permission
> (a) shall not be used as temporary sleeping accommodation as defined in section 25 of the Greater London Council (General Powers) Act 1973, as C amended by the Greater London Council (General Powers) Act 1983*,
> (b) shall not be used for occupation for a holiday whether the right of occupation is created by tenancy or some other form of contract (whether or not such occupation falls within (a) above).

The inspector dealt with condition 2 in paras 17 to 20 of his decision letter. It will be convenient if I recite those paragraphs in full. In para 17 the inspector said:

D

> 17. I now turn to condition No 2 about which you and the Council have made further representations at my invitation, since I called attention to the fact that a condition virtually identical in effect had been considered and discharged by me in a previous decision . . . My conclusion in that decision that part (a) of the condition was unnecessary in the light of section 25 of the Greater London Council (General Powers) Act 1973 was challenged by the Council in the High Court, and the Court upheld my conclusion.
>
> E 18. I shall not repeat here the full reasoning set out in the above-mentioned decision letter, of which you and the Council have copies, but it seems to me that, as regards part (a) of the condition, the circumstances which had a material bearing on my conclusion, and on the court's judgment, in that case are materially the same in this instance. In their further representations the Council have questioned the reasoning of the learned Deputy Judge and maintain that by virtue of the interpretation of the 1973 Act, the condition is F still necessary in order to secure their purpose.
>
> 19. Questions of law are for the Courts and not for me to decide, and it would be inappropriate for me to enter into a debate on the learned Deputy Judge's ruling. However, subject to that proviso, with regard to section 25(1) (*b*) of the Act, I would comment that in the present instance it seems to me there are no grounds for believing that the dwellings are designed or constructed for use as other than permanent residences (notwithstanding that G they may also be suitable for use as "temporary sleeping accommodation") so that, in my view, section 25(1) would apply anyway on that account.
>
> 20. As to part (b) of Condition 2, my view remains that section 25(1) of the GLC (General Powers) Act 1983 (*sic*), by effectively ruling out occupancies of less than 90 days duration, would make this part of the condition also almost superfluous in practice. In so far as the accommodation may perhaps be used by people on holiday for more than 90 days, my opinion is unchanged, that the practical problems entailed in detecting breaches, and H the difficulty of defining the term "holiday" in many instances, would make enforcement extremely difficult and often impossible, and I believe in any case that the process of discovering the reason for which people were living in these dwellings would entail an unacceptable prying into their domestic life. I shall therefore not impose either part of the formal condition No 2 in the new permission.

---

*Editor's note: For full text of section 25, as amended, see p 44 *ante*.

The decision of the High Court to which the inspector was referring was a A
decision of Mr Gerald Moriarty QC, sitting as a deputy judge of the Queen's Bench Division, in the case of *Westminster City Council* v *Secretary of State for the Environment*. That case was heard on June 12 1989. I have been provided with a transcript of the judgment.

Although the notice of motion simply referred to condition 2, Mr Barnes has explained that it was really condition 2(a) that was at issue. As Mr Barnes
put the point, the issue of law before the court is whether condition 2(a) B
confers an element of planning control over and above that conferred by section 25 of the local Act. Mr Barnes contended that it did in three circumstances: first, where a new building or part of a building erected with permission to be used as a residency is first used as temporary residential accommodation. In those circumstances, he said, section 25 does not apply
because of the definition of "residential premises" in section 25(2)(*b*). C

Second, Mr Barnes submitted that, even if section 25 does apply in the above circumstances, no enforcement action against the initial temporary residential use is possible, since it constitutes development that has the benefit of planning permission. Thus, there is no breach of planning control.

Third, it was submitted that, where the building is used initially for permanent residential accommodation and the use is then changed to use as
temporary residential accommodation, that change of use is prevented from D
being development by section 55(2)(f) of the 1990 Act and Class C3 of the Use Classes Order 1987.

He also argued, and there was no dispute on this point, that in each of the above circumstances the use of temporary accommodation as defined in section 25(2)(*a*) would be a breach of condition 2(a) and would be prevented by that condition. Accordingly, Mr Barnes argued that the inspector had
misunderstood the law and the effect of section 25 and wrongly concluded that E
section 25(2)(*a*) was unnecessary.

In the earlier case to which I have referred the same issue was raised and in that case the learned deputy judge concluded that the condition which the local planning authority sought to impose was — and I use his words — "unnecessary because the effect of that condition is produced by section 25 of the Local Act".

Mr Barnes challenges that decision and very properly points out, as appears F
from the judgment, that the argument in that case was put upon a somewhat different basis from that put by Mr Barnes in this case. Mr Barnes specifically challenged the findings of the learned deputy judge as set out at C and E on p 11 (the last page of the judgment) where Mr Moriarty said:

> . . . section 25(1) is indeed a deeming provision and the phrase "involves a
> material change of use" has the effect that each time there is a use as G
> temporary sleeping accommodation of residential premises in Greater London there is a material change of use whether or not the premises have hitherto been used as one or more permanent residences and whether or not they were designed or constructed for use as one or more permanent residences.

On the first of those two points made by the learned deputy judge, Mr
Barnes suggests that, even if there were a change of use involved, there would H
be only one change of use if one temporary use was followed immediately by another.

Mr Sankey, on behalf of the Secretary of State, accepted that, if Mr Barnes' interpretation of what the learned deputy judge was saying was correct, then Mr Barnes' criticism was justified. However, Mr Sankey contended that the error would not invalidate the judgment. With that I agree.

As to Mr Barnes' second point, he urged that to hold that the change of use to which the learned deputy judge referred occurred "whether or not the premises have hitherto been used as one or more permanent residences and whether or not they were designed or constructed for use as one or more permanent residences" was inconsistent with the language of section 25, which should be read as if it said "provided" they had been previously used or "provided" they had been designed or constructed for use as one or more permanent dwellings.

Again Mr Sankey accepted that this last point of the judgment was wrong. For my part I am not entirely clear what the learned deputy judge meant by the language that he used in the last few lines of his judgment, but, be that as it may, Mr Sankey contends that the learned deputy judge was correct in his conclusion on the central issue of the case, namely that the effect of the condition is produced by section 25 of the local Act and that in that regard the judgment correctly stated the law.

Before turning to Mr Barnes' submissions it is right to record that he accepted that if he was correct in those submissions section 25 of the local Act was of very limited effect. In the earlier *Westminster* case the learned deputy judge reached the same conclusion and said in terms that he was influenced by the fact that if the local planning authority's view was right section 25 seemed to have a very limited effect. I have sympathy with the learned deputy judge's view upon that matter.

I turn, then, to Mr Barnes' three contentions as they are very helpfully set out in para 4 of the skeleton argument. I propose to deal with them seriatim.

As I have already indicated, Mr Barnes' first submission was that, where a new building or part of a building was erected with permission to be used as a residence and is first used as temporary accommodation, section 25 does not apply because of the definition of "residential premises" in section 25(2)(*b*). I must confess in passing that section 25 might be much easier to understand if subsection (*b*) were not included and that the omission of subsection (*b*) would not seem to affect the potential operation of the section. However, it is there and must be addressed.

In developing his argument upon this matter Mr Barnes postulated the position where the grant of consent was neutral as to whether the premises were intended to be used as temporary accommodation as defined in section 25(2)(*a*) or on what might be regarded as a more normal and longer-term basis. Mr Barnes — who urged that the word "designed" in section 25(2)(*b*) means "intended to be used" rather than "of an architectural design suitable for a particular use" — submitted that in those circumstances it would be impossible to say that the building was designed or constructed for one rather than another of the purposes to which I have referred. Mr Barnes further submitted that if someone has what he described as a "neutral" planning permission for residential development they could develop with the intention to use the premises either on a temporary basis or on a longer-term basis. Certainly, if it were not for section 25, although I do not intend at the moment to deal with the effect of that section, that would appear to be the position. Consequently, Mr Barnes urged that a first user on a temporary basis would not require any further planning permission in the example that he put forward and that user could not be prevented.

In making this submission he placed reliance upon the case of *Wilson* v *West Sussex County Council* [1963] 2 QB 764. In that case outline permission was granted for the erection of an agricultural cottage in accordance with the applications and plans submitted. Permission was granted subject to certain conditions. In particular, Mr Barnes referred to passages in the judgments of

A Diplock LJ and Willmer LJ. At p 777 Willmer LJ expressed himself as follows:

> In the circumstances I think it is impossible to avoid the conclusion that it is proper, and indeed necessary, in construing these planning permissions to refer to the terms of the applications. That being so, in the context of these applications, what was meant by the phrase "agricultural cottage"? We are not, it is to be observed, construing the words of a statute: we are construing words which, as I have shown, appear to have had their origin in a phrase B chosen by the original applicant in the first instance. They must, in the circumstances, as it seems to me, be construed in their natural and ordinary sense. I think there can be no doubt that Mr Ryan was well founded when he submitted that the word "agricultural" in this context must be construed as having a functional significance.

Willmer LJ went on to conclude that the phrase must be construed as C limiting the user of the building that is proposed to be erected.

At p 783 Diplock LJ, referring to that same phrase, which, as is clear from Willmer LJ's judgment, appeared in the application and was incorporated within the grant, said:

> But in any event it seems to me that it defines the purpose for which the cottage is designed, for "designed" in the context of an outline permission must mean "intended to be used". It cannot mean "of an architectural D design suitable for a particular use", for at the time of the outline permission there may not be in existence, and there were not in this case any architectural designs for the cottage at all, and "designed" in the subsection cannot have a different meaning according to whether the permission is an outline permission granted before architectural designs are in existence or a final permission granted after architectural designs have been prepared.

Mr Sankey does not accept that the word "designed" in section 25(2)(*b*) E should be interpreted in the present case in the way that it was interpreted with regard to different sections of a different Act in *Wilson*. In *Wilson* the decision was made in relation to section 18(3) of the Town and Country Planning Act 1947, which is now section 75(2) and (3) of the 1990 Act.

There are, in my judgment, certain significant differences between the position in *Wilson* and the position in this case. In this case the question of F enforcement, which is one of Mr Barnes' concerns, can arise only when the premises have been built and are in fact being or have been used. It is upon that basis that Mr Sankey suggests that the proper way to interpret the definition of "residential premises" in section 25(2)(*b*) is to ask whether the premises are designed or constructed so as to be capable of use as a permanent residence. If that is the right approach, then Mr Barnes accepts that his first point goes.

I have already indicated that I see differences in the circumstances between G this case and *Wilson*, and I am inclined to the view that Mr Sankey's submission is correct. But even if I were not inclined to that view, I would reach the same conclusion in terms of the validity of Mr Barnes' point even if I were to follow the *Wilson* approach. I do not believe that it could be suggested, looking at the *Wilson* example, if somebody obtained a planning permission by way of a grant for residential use or a residential development *simpliciter* H and then happened first to use it in connection with agriculture, that the permission would be limited by the fact of that use — even if it had been the developer or owner's intention so to use it from the outset. *Wilson*, in my view, turned upon the meaning and effect of the words "agricultural cottage", which were, as the learned lord justices pointed out, incorporated in fact in the grant by reference to the application — that was the key part. It was those words that limited what would otherwise have been a normal and full residential permission.

A Turning, then, to Mr Barnes' submission, it seems to me that it would not be right to investigate what might have been in the mind of the potential developer in the absence of a limiting phrase or a definitive phrase in the grant, albeit it might be incorporated as a result of something said in the application which was itself translated into the grant. What is granted when permission is granted for residential development is permission to use for residential purposes.

B At first glance that might seem to support Mr Barnes' contention, since the premises can be used on the face of it in the course of the ordinary law for any residential purpose, including a temporary use as defined in section 25(2)(*a*). But such a conclusion is to deny the effect and presence of section 25(2)(*a*), and in this regard, whatever criticism may be made of the final words of the learned deputy judge, Mr Gerald Moriarty QC, in his judgment in the earlier
C *Westminster* case, I agree that section 25(1) is a deeming provision and that the phrase "involves a material change of use" has the effect, in the example postulated by Mr Barnes in his first submission, that in the absence of some specific reference in the grant of permission to the right to use the premises for temporary purposes it is unlawful so to use them even on a first use in the area to which section 25 applies.

I turn now to Mr Barnes' second submission, which was to the effect that,
D even if section 25 does apply in the above circumstances, no enforcement action against the initial temporary residential use is possible, since it constitutes development that has the benefit of the planning permission. Thus, says Mr Barnes, there is no breach of planning control.

I have already effectively dealt with that point. In the absence of a specific grant for use as temporary sleeping accommodation within the definition in section 25(2)(*a*), there would be, in the case of premises to which the section
E applied, a breach of planning control and it would be possible subject to the third submission, although I am not sure that that actually turns upon the point, to enforce against that use.

The third point taken by Mr Barnes, as I have already indicated, ran as follows. Where the building is used initially as permanent residential accommodation and the use is then changed to use as temporary residential
F accommodation, that change of use is prevented from being development by section 55(2)(f) of the 1990 Act and Class C3 of the Use Classes Order 1987. Section 55(2)(f) provides as follows:

> The following operations or uses of land shall not be taken for the purposes of this Act to involve development of the land —
>
> . . .
>
> (f) in the case of buildings or other land which are used for a purpose of any
> G class specified in an order made by the Secretary of State under this section, the use of the buildings or other land or, subject to the provisions of the order, of any part of the buildings or the other land, for any other purposes of the same class.

Class C3 of the Use Classes Order 1987 provides as follows:

> Use as a dwellinghouse (whether or not as a sole or main residence) —
> H (a) by a single person or by people living together as a family, or
> (b) by not more than 6 residents living together as a single household (including a household where care is provided for residents).

That use class was a new class introduced in the 1987 order and, certainly unlike a number of the other use classes, does not carry within it more than one use. It is simply use as a dwellinghouse. The purpose of that new use class I think is clearly understood and was to cover problems that had previously occurred as to whether in certain circumstances there was a change of use if a dwellinghouse was used, for instance, in multi-occupancy.

But Mr Sankey submits that the significant thing about Class C3 in the light of the provisions of section 52(2)(f) of the Act is that, since it is a one use class, there cannot be a change for any other purpose within the class. I accept that submission. It seems to me that the position that emerges here is that but for section 25(2)(*b*) planning permission would not in any case be required to use residential premises for temporary accommodation as defined in section 25(2)(*a*), and that would not be by virtue of the provisions of Class C3 but simply because it would still be used as a dwellinghouse.

In the circumstances, my judgment is that the effect of section 25 is that what would otherwise have not required planning permission requires planning permission, and therefore to change from an initial use as permanent residential accommodation does, by virtue of that section, require planning consent. To change without that consent having been granted is unlawful and capable of enforcement.

It follows that, in my judgment, the central finding made by Mr Gerald Moriarty is correct and, to use his words, "the condition is unnecessary", as the inspector found, because the effect of that condition is produced by section 25 of the local Act.

I should just add this with regard to the first of Mr Barnes' submissions, and that is that in the decision letter the inspector concluded in para 19 that in the case before him there were no grounds for believing that the dwellings were designed or constructed for use as other than permanent residences, so he said that section 25(1) would apply anyway on that account. I do not see any basis upon which that particular view can be criticised and that reinforces the view that I have reached in this judgment.

I am not sure whether Mr Barnes or his clients are pleased or discontented with this judgment, since Mr Barnes made it plain at the outset that it was a somewhat unusual case and the authority either needed the condition or at least the comfort of knowing that it was not needed rather than simply to have this decision quashed. But, be that as it may, this application is dismissed.

Application dismissed with costs.

Bath Society *v* Secretary of State for the Environment and others

COURT OF APPEAL
Glidewell and Stocker LJJ and Sir Denys Buckley
February 6 1991

Material considerations — Conservation area — Statutory duty of decision-maker — Recommendations of local plan inspector concerning open space — Whether a material consideration — Whether knowledge of Secretary of State — Appeal inquiry inspector — Whether Secretary of State ought to have considered recommendations of local plan inquiry inspector — Duties under sections 29(1) and 277(8) — Principles for fulfilling statutory duty relating to conservation area set out

By a decision dated November 25 1988 the first respondent, the Secretary of State for the Environment, allowed an appeal by the third respondents, Hammercrest Developments Ltd, against a refusal by Bath City Council of an

A application for planning permission for the construction of a block of flats and garages. The subject site forms the garden and curtilage of a former house, Cavendish Lodge. After the hearing of the appeal, but before the first respondent's decision, an inspector holding an inquiry into the draft local plan made recommendations in relation to the site. The appellant society appealed against the decision of Hutchison J (February 19 1990), who had dismissed their application to quash the first respondent's decision. On appeal they
B submitted that: (1) the first respondent had failed to take into account the recommendations of the inspector appointed in relation to an inquiry into the local plan that the subject site should be allocated as open space, alternatively he had failed to defer a decision until the city council's response to the inspector's recommendation was known; and (2) the Secretary of State had failed to comply with the duty imposed on him by section 277(8) of the Town and Country Planning Act 1971.

C

Held The appeal was allowed and the first respondent's decision was quashed.

The recommendation of the local plan inquiry inspector was a material consideration: see *In re Findlay* [1985] AC 318. His report was known to the first respondent, who was in breach of his statutory duty in failing to have regard to the recommendations: see p 59.

D It was agreed that the duty imposed by section 277(8) of the 1971 Act (section 72(1) of the Planning (Listed Buildings and Conservation Areas) Act 1990) applied. Where permission is sought for development on a site not within a conservation area, the policy in para 15 of PPG 1 is a perfectly proper approach to the carrying out of the statutory duty under section 29(1) of the 1971 Act. If the site is within a conservation area, the decision-maker has two statutory duties imposed on him: sections 29(1) and 277(8). The requirement in section 277(8) to pay "special attention" to the desirability of preserving or
E enhancing the character or appearance of a conservation area should be the first consideration of the decision-maker; it is a material consideration of considerable importance and weight. If the development will enhance or preserve the character of the conservation area, this must be a major point in favour of allowing the development. In some cases a development may simultaneously enhance the character or appearance of the area but nevertheless cause some detriment. The detrimental effect is a material
F consideration to be weighed against the contrary consideration. If the decision-maker concludes that the proposed development will neither preserve nor enhance the conservation area the presumption derived from para 15 in PPG 1 is then rebutted: see p 64E *et seq*.

In the present case the inspector did not comply with the section 277(8) duty; if he had, and formed the opinion that the building would neither enhance nor preserve the character or appearance of the conservation area, then he failed to carry out the necessary balancing exercise of weighing that important
G material consideration against such unexpressed benefits as the construction of the new building might bring: see pp 65-66.

Cases referred to in the judgments

Creednz Inc v *Governor-General* [1981] 1 NZLR 172

Findlay, In re [1985] AC 318; [1984] 3 WLR 1159; [1984] 3 All ER 801, HL

Harrow London Borough Council v *Secretary of State for the Environment* (1989) 60
H P&CR 525; [1990] 2 PLR 62

Hollis v *Secretary of State for the Environment* (1982) 47 P&CR 351; [1983] EGD 1046; 265 EG 476; [1983] JPL 164

South Lakeland District Council v *Secretary of State for the Environment* unreported February 26 1990; on appeal [1991] 2 PLR 97, CA

Steinberg v *Secretary of State for the Environment* (1989) 58 P&CR 453; [1989] 2 PLR 9; [1989] JPL 258

Unex Dumpton Ltd v *Secretary of State for the Environment* [1990] 2 PLR 1; [1990] JPL 344

Ward v *Secretary of State for the Environment* (1989) 59 P&CR 486; [1990] 1 PLR 85, CA

Wyre Forest District Council v *Secretary of State for the Environment* (1989) 58 P&CR 291; [1989] 2 PLR 85; [1989] JPL 362, CA

Appeal against decision of Hutchison J

This was an appeal against the decision of Hutchison J (February 19 1990), who had dismissed the appellant society's application under section 245 of the Town and Country Planning Act 1971 (section 288 of the Town and Country Planning Act 1990) to quash the decision of the first respondent, the Secretary of State for the Environment, who had allowed an appeal by the third respondents, Hammercrest Developments Ltd, against the decision of the second respondents, Bath City Council, to refuse planning permission.

Major A J W Crombie appeared, with the leave of the court, for the appellants, the Bath Society.

John Howell and Alison Foster (instructed by the Treasury Solicitor) appeared for the first respondent, the Secretary of State for the Environment.

Anthony Dinkin (instructed by Portner & Jaskel) appeared for the third respondents, Hammercrest Developments Ltd.

The second respondents, Bath City Council, did not appear and were not represented.

Cur adv vult

The following judgments were delivered.

GLIDEWELL LJ: On September 9 1987 Bath City Council, the local planning authority, refused an application by Hammercrest Developments Ltd for planning permission to construct a four-storied block of 20 flats, two lodges, garages and roadways on land adjoining the site of Cavendish Lodge, Lansdown, Bath. Hammercrest appealed. Their appeal was heard by an inspector, Mr MacDonald, an architect, in May 1988. In July 1988, he reported to the Secretary of State for the Environment, recommending that the appeal should be allowed. The Secretary of State accepted this recommendation and, by a decision letter dated November 25 1988, allowed Hammercrest's appeal.

At the inquiry at which the appeal was heard, the Bath Society were represented by Major Crombie, who on their behalf strongly objected to the proposed development. On receiving the Secretary of State's decision, the society sought to challenge it by way of an application to the High Court under section 245 of the Town and Country Planning Act 1971. On February 19 1990, after a four-day hearing in Plymouth, Hutchison J dismissed the society's application. The society now appeal to this court.

Before us the society are again represented by Major Crombie, who has (no doubt with the assistance of other members of the society) prepared their case with great care and competence, provided an admirable skeleton argument, and addressed us with skill and courtesy. The court is much indebted to Major Crombie for his efforts, as indeed are the other members of the Bath Society.

The site and its surroundings

Bath is a city of great beauty, with a wealth of historic buildings constructed for the most part in what Sir Arthur Bryant called "the age of elegance", ie the 18th and early 19th centuries. It is among the sites and cities listed by the World Heritage Convention administered by UNESCO as "of such exceptional interest and of such universal value that protecting them is a concern of all mankind". The city rises northwards from the valley of the River Avon. Much of the Georgian city consists of terraces of stone houses interspersed with open spaces and gardens. The appeal site lies close to two of these terraces, being to the east of Cavendish Crescent and to the south of Lansdown Place. The site forms the garden and curtilage of a former house, Cavendish Lodge. The site

A itself is a T-shaped area of open land, 2.7 acres in extent, comprising what is called "the garden" (the stem of the T) and the "field". The development proposed is the construction of a block of flats and garages on the field, with lodges on the garden at the entrance to the entire development. There are a number of trees subject to tree preservation orders on the garden.

Planning history

B The planning history of this site is extensive. I cite only that part of it which is relevant for present purposes.

On February 4 1964 (before the conservation area came into being) outline permission was given for the construction of up to 30 residential units on 1.7 acres forming part of the site. This was described as being "east of Cavendish Lodge", which was then standing. There was no time-limit within which this development was to be carried out. The permission specifically included C consent for a departure from the development plan.

On December 5 1967 outline permission was granted for the construction of 36 flats and 12 maisonettes on the field.

In December 1968 six conservation areas, one of which included the appeal site, were designated by the city council under the Civic Amenities Act 1967. These conservation areas were later amalgamated into one.

On March 28 1972 an application was made for approval of details under D the 1967 outline permission. By virtue of Schedule 24, para 20(1), to the 1971 Act, the outline permission remained effective provided that such an application was made before April 1 1972. In other words, the application was in time and the planning permission remained in force. In addition, in May 1972 a new application was made for planning permission to erect 48 flats on the field. The city council did not decide the application for approval of details and refused the application for 48 flats. An appeal was entered against both E the failure to decide the first application and the refusal of the second application.

This appeal was heard at an inquiry in March 1973. By a decision letter dated April 22 1974 the Secretary of State dismissed both appeals. I shall return later to the reasons for these decisions.

Listed building consent to demolish Cavendish Lodge was granted in July 1982 and the house was demolished thereafter.

F A development brief for the appeal site was approved by the city council on May 6 1986. This brief contained the following provisions among others:

> (a) No new buildings, other than a lodge, are to be erected significantly in front of 11 Cavendish Crescent and they may not project back into the field. Garages and/or parking spaces together with not more than three dwellings would be acceptable at the southern end of the field.

G This restricted building development to two relatively confined areas of the site.

> (b) Landscaping. None of the trees on the site are protected by the Tree Preservation Order.
>
> (c) To safeguard the field against future development, before planning permission is granted, the owners of the property will be required to enter into an agreement under section 52 of the Town and Country Planning Act 1971 to secure the laying out and retention of the field as an amenity space.

H *The appeal*

By letter dated March 31 1988, before the appeal hearing, Hammercrest's architect submitted amended plans. The inspector agreed to accept these for the purposes of the appeal, concluded that they were preferable to the plans originally submitted, and based his recommendation upon the amended plans. They show a design for a block of flats in what I would call an ornate classical style. Of it the inspector said in his report:

> The design of the proposed building has been criticised for being incorrect, A
> within the grammar of classical design or, at least, idiosyncratic in its use of classical elements. The architect has certainly used the elements of the classical vocabulary in a free, and in some respects, non-traditional way, but the resulting building is not unpleasing nor out of character with its setting.

He concluded that:

> . . . the proposed building would be appropriate to its site and setting. It B
> represents a successful attempt to design in harmony with classical principles, but without slavish adherence to precedent.

Before Hutchison J Major Crombie based one of his submissions on an objection to the inspector's acceptance of the amended plans, but he has not repeated this submission before us. Nothing, therefore, now turns on this matter. C

The Secretary of State's reasons for allowing the appeal were expressed in the following words:

> The Inspector recommended that, in view of the considerations expressed in paragraphs 42 to 45 of his report, the appeal should be allowed, subject to conditions. The Secretary of State agrees with the Inspector's conclusions and accepts his recommendation.

It follows, therefore, that the challenge to the Secretary of State's decision D
is, in effect, a challenge to the reasoning contained in the inspector's report, and in particular to para 42. Before I come to this, however, there is one other matter with which I should deal shortly.

The local plan

Before the appeal hearing Bath City Council, as local planning authority, had prepared and published a local plan for the city. An inquiry into objections to E
this plan was heard in April 1988 by an inspector, Mr Gray. Local plan inquiries are heard by inspectors of the Department of the Environment inspectorate, but they are in effect on loan to the relevant local planning authority. The inspector who holds such an inquiry reports to the local planning authority, not to the Secretary of State. It is then for the local planning authority to decide whether and how far to adopt his F
recommendations, to advertise any modifications to the plan, and if necessary to hold a further inquiry or to consider representations about objections to these modifications.

In the local plan, the garden part of the appeal site was allocated for residential development, whereas the field was shown with no notation. The city council explained that this was to be read in conjunction with the development brief, to which I have already referred. The Bath Preservation G
Trust objected to the lack of notation on the field, urging that it should be shown as open space. They made this objection at the inquiry into the plan.

Post-inquiry history

After the hearing of the appeal, on June 28 1988 Mr Gray reported to the council on his inquiry into the local plan. He set out the objection by the Bath Preservation Trust, the council's response to it, and concluded: H

> 8.37 The land to the east of Cavendish Lodge referred to by the Bath Preservation Trust is clearly an open space necessary to the setting and character of buildings in the locality. This is recognised in the Planning Brief prepared by the Council and as a consequence I see no reason for the omission of this land from the sites illustrated on Map 8 [as open space].

His formal recommendation was:

A
> That the Cavendish Lodge field be shown as an open space as recommended in paragraph 8.41.

In July 1988, as I have said, Mr MacDonald, the appeal inspector, sent his report to the Department of the Environment.

On August 30 1988 the council sent a copy of Mr Gray's report on the local plan inquiry to Mr Rollinson, the regional superintending planner in the
B south-west regional office of the Department of the Environment. Mr Rollinson had asked to be supplied with a copy of Mr Gray's report for purposes wholly unconnected with the planning appeal in respect of the Cavendish Lodge site. He was, however, concerned in the process of making the decision on that appeal. Mr Rollinson, after considering parts of Mr Gray's report, passed it on to another colleague, but as I understand it it remained in the regional office of the Department.

C On November 25 1988 the Secretary of State issued his decision letter on the appeal.

On December 20 1988 the council accepted a recommendation of their relevant committees to adopt the recommendation of the local plan inspector in regard to the Cavendish Lodge site. On January 23 1989 the council issued a list of modifications to the local plan, one of which showed the field as open space.

D

The appeal inspector's reasoning

In standard form, the appeal inspector's report contained findings of fact and then conclusions. His findings of fact included:

> The emerging Local Plan, the draft City Plan, shows the "garden" as a site for residential development, but leaves the "field" unannotated.

E In his next finding he summarised the residential development brief. He noted that:

> The appeal site has been included in a Conservation Area since December 1968.

The part of his conclusions which is relevant to this appeal is para 42, which it is necessary to set out in full. It reads as follows:

F

> 42. The most contentious aspect of the appeal proposals is the proposed location of the main buildings on the eastern part of the site, ie on the "field". It is this aspect to which the amenities societies fundamentally object, as do many of the interested persons, and it is in this respect that the proposal departs most significantly from the City Council's approved Development Brief. However, it must be noted that the Secretary of State, in his 1974
> G decision letter, accepted that development of this part of the site could reasonably be allowed, and that the appeal proposals were dismissed solely on design considerations. The desire to see the "field" become publicly-owned open space is understandable, but it is clear that no advance in this direction has been made after many years of discussion, and no firm proposals were put forward at the Inquiry for moving towards this objective. It is also clear that the argument for preserving the field was put forward at the 1973 inquiry, but did not persuade either the Inspector or the Secretary of
> H State, and that use as a public open space is no nearer now than it was then. Notwithstanding the provisions of the Development Brief, there are cogent reasons for locating a new building in the position proposed by the appellants, in particular that it removes the building from that area of the site where 8 of the 9 protected trees are located, and sets the building behind the gable end of Cavendish Crescent, thus minimising the visual impact of the proposed building on this important terrace. The proposed location appears, when viewed on site, to be the natural siting for any major new building.

> Because of the fall of the site from north to south, the roofline of the new building would be below the eye-level of a person standing in Lansdown Crescent West, and its effect on the view looking southwards from there would not be unacceptably great. The effect on this view would, of course, be very much less if any new building was sited upon the "garden" area, but while the visual impact upon Lansdown Crescent would be lessened, that upon Cavendish Crescent would, I believe, inevitably be much increased. Weighing these issues, I find myself in agreement with the principle set out in the 1974 decision letter, that there is no fundamental objection to the siting of a suitably designed building on the "field".

The following paragraphs of the inspector's conclusion deal with other issues, ie para 43 with the detailed design of the building, para 44 with the retention of trees, and para 45 with conditions.

Submissions

Before Hutchison J Major Crombie made submissions on eight grounds. Before us he has restricted himself to two of these, namely:

(1) The failure by the Secretary of State to take into account the recommendation of Mr Gray, the local plan inquiry inspector, that the field should be allocated as open space, or alternatively to defer a decision on the planning appeal until that recommendation, and the council's response to it, were known to the Secretary of State.

(2) A failure by the Secretary of State to comply with the duty imposed on him under section 277(8) of the 1971 Act.

I will consider each of these in turn.

(1) Major Crombie's argument on the first of these two issues can be summarised as follows:

(a) The recommendation of the inspector who held the local plan inquiry was clearly a relevant consideration which the Secretary of State should have taken into account when making his decision on the planning appeal.

(b) The inspector who held the appeal inquiry knew of the Bath Preservation Trust's objection to the failure to allocate the field as open space in the local plan. He reported to the Secretary of State that there was an objection but did not give any detail of it. Moreover, on August 31 1990 Mr Rollinson, who was concerned in the decision-making process on the appeal, was sent a copy of the local plan inquiry inspector's report, including the recommendation to allocate the field as open space.

(c) It follows, therefore, that the Secretary of State must be deemed to have known of the open space recommendation, but disregarded it, and so failed to have regard to a material consideration.

(d) Alternatively, the Secretary of State, through his officers, should have known that the local plan inquiry inspector would be making a recommendation on this issue and should have taken steps to ascertain what that recommendation was before making his decision on the planning appeal, if necessary by deferring it. In failing to take steps to make himself acquainted with the recommendation he acted irrationally.

If I understand him correctly, Mr Howell, for the Secretary of State, is inclined to argue that the local plan inquiry inspector's recommendation was not necessarily a material consideration. The decision whether it was material was a decision on a question of fact, to be made by the Secretary of State. Such a decision can be challenged only on ordinary *Wednesbury* principles.

With respect to Mr Howell, it seems to me that this argument misses the point. Major Crombie is not complaining that the Secretary of State considered whether the recommendation on the local plan was or was not material and reached a wrong conclusion; part of his complaint is that there is

A no evidence that anybody on behalf of the Secretary of State considered the matter at all. If this be correct, as in my view it is, then it is open to the court to decide whether the recommendation was a material consideration. It was a recommendation which, if accepted by the council (as it now has been), could lead to the local plan containing an allocation of the site as open space with which the proposed development would be totally inconsistent. *In re Findlay* [1985] AC 318 Lord Scarman, in a speech with which the other members of the
B House agreed, quoted with approval two short passages from the judgment of Cooke J in *Creednz Inc* v *Governor-General* [1981] 1 NZLR 172. The second passage included the observation that, though the relevant statute does not refer expressly to a particular consideration, nevertheless "there will be some matters so obviously material to a decision on a particular project that anything short of direct consideration of them by the ministers . . . would not
C be in accordance with the intention of the Act": see [1985] AC 318 at p 334B.

In my view, the recommendation of the local plan inspector came within this description.

The combined effect of sections 29(1) and 36(5) of the Act of 1971 is that the Secretary of State, in deciding whether to allow an appeal under section 36, is required to have regard to the provisions of the development plan and to any other material considerations. So it appears that, if the view I have expressed
D above is correct, the Secretary of State failed to comply with the statutory requirement. Nevertheless, from that fact alone, the court is not *obliged* to quash his decision — it *may* do so, in its discretion. In exercising that discretion, a relevant question for the court is whether the Secretary of State knew, or ought reasonably to have known, of the matter which was material to his decision. It was, in effect, upon his answer to this question that Hutchison
E J based his decision on this issue. He said (at p 19A of the transcript of his judgment):

> Mr Laws [counsel for the first respondent, the Secretary of State for the Environment], in answering these submissions, did not invite me to embark upon a consideration of the concept of the "indivisible crown". Of the submissions he made I propose to mention only two. The first was that, given the circumstances in which Mr Rollinson received the report in August and
> F the fact that Mr Rollinson was entirely unconnected with the Cavendish Lodge appeal, Mr Rollinson would not have been free to circulate the report to the decision-making officers: for they would have had to publicise it and that would have been contrary to the confidence in which it had been sent by the council. This, as it seems to me, is only a partial answer to Major Crombie's argument, for if it be right to assume that the decision-making officers have notional knowledge of the contents of the local plan inspector's
> G report, then breaches of confidence would have been averted by postponing a decision until after the local authority's response to the local plan inspector's report had been made public.
>
> However, the second argument does appear to me to be a conclusive answer to the applicants' case. It is a simple argument on the facts, and proceeds thus. It was known on all sides that the local plan inspector was holding an inquiry, and that that inquiry might well touch upon this site. Before the appeal inspector no request was made either by the council or by
> H the Bath Society or by anybody else for an adjournment, or for the inspector to postpone his report until the result of the local plan inspector's inquiry was known. The Bath City Council, in possession of the local plan inspector's recommendations, and presumably better acquainted than anyone else with their likely response to it, made no representations to the Secretary of State, nor did they request him to postpone his decision on the appeal until they had formally resolved what to do about the local plan inspector's report. In the circumstances, submits Mr Laws, what can be said to be unreasonable or

> unfair about that: in what respect can it be said that there has been a failure to take material matters into consideration? A

The learned judge concluded:

> As I have said, no application for deferment or an adjournment was made and the plan in the possession of the local planning authority was not referred to by them or sent to the Secretary of State. I accept the arguments advanced by Mr Laws and Mr Dinkin that in the circumstances of this case there are no valid grounds for criticising the inspector or the Secretary of State in regard to their attitude towards the emerging local plan. B

Major Crombie has referred us, as he did Hutchison J, to *Hollis* v *Secretary of State for the Environment* (1982) 47 P&CR 351, a decision of mine at first instance quashing a decision of the Secretary of State on the ground that there were in the relevant part of his department, and thus within his knowledge, C documents which made it clear that information recorded in an inspector's report, on which the decision was based, was incorrect. He relies upon this particularly because, as Mr Howell concedes, Hutchison J misunderstood Mr Rollinson's position when he said that "Mr Rollinson was entirely unconnected with the Cavendish Lodge appeal". Since Mr Rollinson was concerned in the process of deciding the appeal, the information he possessed, submits Major Crombie, was that of the Secretary of State, and he had (at least D for some time) in his possession the local plan inspector's report.

Mr Howell refers us to the decision in *Wyre Forest District Council* v *Secretary of State for the Environment* [1989] 2 PLR 85. In that case development, for which planning permission had been refused by the district council, was inconsistent with the zoning in the local plan, which was to be the subject of an inquiry at approximately the same time as the hearing of the E developers' appeal against the refusal of permission. The inspector hearing the appeal was asked, but refused, to defer making a decision until the local plan inspector reported. This court held that this was a proper exercise of his discretion. But, as I have said, in the present case the Secretary of State was not asked to defer his decision and there is no evidence that he considered doing so.

I do not find the decision in *Hollis*, which was made on different and unusual facts, of much assistance in answering the question before us. On the F one hand, it would have been a simple matter for the council to draw the attention of Mr Rollinson, or somebody else in the decision-making branch of the Department of the Environment, to the "open-space" recommendation, and perhaps to ask for the appeal decision to be deferred until the council had decided whether to accept the recommendation. The fact that the council failed to do so is not the fault of the Secretary of State. On the other hand, the appeal inspector did report that there had been an objection relating to the G field at the local plan inquiry, and it would have been equally easy for Mr Rollinson, or one of his colleagues, to have studied the local plan inspector's report to ascertain what he recommended in relation to this site and then to take that recommendation into account in some way.

On balance I conclude that the failure of those who were deciding the appeal to take these simple steps resulted in the Secretary of State's failing to comply with his duty to have regard to this material consideration. H

(2) I now turn to Major Crombie's second submission, namely that the Secretary of State is not shown to have complied with the duty imposed on him by section 277(8) of the 1971 Act.

Section 277 contains the following relevant provisions:

> **277.**—(1) Every local planning authority shall from time to time determine which parts of their area are areas of special architectural or historic interest

> the character or appearance of which it is desirable to preserve or enhance, and shall designate such areas as conservation areas.
>
> . . .
>
> (8) Where any area is for the time being designated as a conservation area, special attention shall be paid to the desirability of preserving or enhancing its character or appearance in the exercise, with respect to any buildings or other land in that area, of any powers under this Act, . . .

There is general agreement that the duty imposed by section 277(8) applied to the exercise by the Secretary of State of his powers when considering the appeal to him under section 36 of the 1971 Act.

Neither the Secretary of State in his decision letter nor the appeal inspector in his report says expressly that he has paid attention to the desirability of preserving or enhancing the character or appearance of the conservation area. Thus, submits Major Crombie, Hutchison J should have found that they did not comply with their duty under the subsection. The learned judge, however, concluded that various passages in the inspector's report (to which I shall have to refer later) "inevitably lead to the inference that the inspector was alive to and was giving proper weight to his section 277(8) duty". The issue can thus be shortly expressed — was the learned judge correct in drawing such an inference?

Major Crombie accepts that in an appropriate case it may be clear, although there are no express words, that an inspector has complied with the statutory requirement imposed by section 277(8). But he argues that here it is apparent that the inspector did not do so, not merely because he has not referred to the section 277(8) duty in terms, but for the following additional reasons:

(i) in para 42 of his report the inspector did consider the effect of the new building on the view and decided that it would not be "unacceptably great". Major Crombie characterises this as a reference to the "harm test" as distinct from the section 277(8) requirement;

(ii) the inspector has placed reliance upon, but misinterpreted, the 1974 decision;

(iii) the inspector adopted a wrong test in comparing the impact of a large building on the field as opposed to the same building in the garden.

I will consider the second and third of these arguments in more detail before returning to the first.

The 1974 decision

The relevant passage is to be found in para 5 and part of para 6 of the 1974 decision letter. This read as follows:

> 5. The Inspector recommended that the appeal against the failure to issue a decision within the prescribed period on the application for approval of matters reserved in the outline permission granted in December 1967 be allowed unless, in the meantime, the outline permission was revoked or modified. He also recommended that unless the permission granted on 5 December 1967 should in the meantime be revoked or modified, the appeal against the refusal of planning permission for the erection of a block of 48 flats and 53 garages on land at Cavendish Lodge and the rear of Cavendish Road be allowed. The Inspector made no recommendation in the alternative event, in the absence of fuller evidence.
>
> 6. The Secretary of State accepts the Inspector's findings of fact and does not disagree with him on any issue of fact. In respect to the Inspector's conclusions, the Secretary of State agrees that some form of new residential development could reasonably be allowed on the appeal site provided the buildings harmonise with the existing buildings and make a worthy contribution to the architecture of this outstanding conservation area.

The Secretary of State then considered the detail of the two applications and A
refused both appeals on the ground that they "do not adequately harmonise with the surrounding architecture nor reach the high architectural quality required in this location".

Major Crombie argues that, although on those appeals the inspector recommended that the appeals should be allowed if the 1967 provision were not revoked or modified, it is clear that his recommendation was upon the
hypothesis that there was an existing planning permission in being. That B
hypothesis is equally obviously the basis of the Secretary of State's agreement that "some form of new residential development could reasonably be allowed on the appeal site". The appeal inspector, Mr MacDonald, in his report (the passage at the beginning of para 42) appears not to have understood this or at least has disregarded it. To this, Mr Howell, for the Secretary of State, replies
that Mr MacDonald not merely considered the 1974 decision but expressly C
exercised his own judgment in concluding, at the end of para 42, that "I find myself in agreement with the principle set out in the 1974 decision letter, that there is no fundamental objection to the siting of a suitably designed building on the 'field' ".

The comparison of the field and the garden as a suitable site for the proposed building

The inspector made this comparison when he said, in the middle of para 42: D

> Notwithstanding the provisions of the Development Brief, there are cogent reasons for locating a new building in the position proposed by the appellants, in particular that it removes the building from that area of the site where 8 of the 9 protected trees are located, and sets the building behind the gable end of Cavendish Crescent, thus minimising the visual impact of the proposed building on this important terrace.

E

Major Crombie submits that the restrictions on development of the site set out in the development brief would result in a relatively modest amount of building on the garden, since it was a requirement of the brief that the buldings should be set back and the trees retained. Thus, it would be impossible to erect the building here proposed on the garden and comply with the development brief. The comparison which the inspector made in the passage I have just
quoted was therefore invalid and should not have formed any part of his F
reasoning.

The section 277(8) requirement

We were referred to five previous decisions of the courts in which the proper application and effect of section 277(8) have been considered. Four of these were decisions of distinguished Queen's Counsel, sitting as deputy High Court
judges, and one a decision of this court. I will consider them in chronological G
order.

In *Steinberg* v *Secretary of State for the Environment* [1989] 2 PLR 9 Mr Lionel Read QC gave judgment on November 25 1988. Camden London Borough Council refused permission for the erection of a single house in a small conservation area. An inspector allowed the developer's appeal against this decision. In his decision letter he said that the second main issue to be
decided was "whether the proposed development would harm the character of H
the conservation area". He concluded that the proposed house was "an acceptable infill" and therefore allowed the appeal. On a challenge under section 245 by Professor Steinberg, a neighbour, the deputy judge held that the inspector had adopted the wrong test. He said at p 12F:

> There is, in my judgment, a world of difference between the issue which the inspector defined for himself — whether the proposed development would

> "harm" the character of the conservation area — and the need to pay special attention to the desirabilty of preserving or enhancing the character or appearance of the conservation area. In short, harm is one thing; preservation or enhancement is another. No doubt the inspector has demonstrated his concern that the character of the conservation area should not be harmed. That, in my judgment, is not the same as paying special attention to the desirability of preserving or enhancing that character as well as its appearance. The concept of avoiding harm is essentially negative. The underlying purpose of section 277(8) seems to me to be essentially positive.

Mr Howell, for the Secretary of State, criticises this passage as too sweeping. He submits that a conclusion that a particular development will not harm a conservation area is equivalent to finding that it will preserve the character or appearance of that area.

Unex Dumpton Ltd v *Secretary of State for the Environment* [1990] 2 PLR 1 was a decision of Mr Roy Vandermeer QC given on July 12 1989. An application for outline permission for residential development on a seven-acre site at Newbury was refused and an appeal against the refusal was dismissed by an inspector. The inspector had posed himself the question whether the proposed development would contribute to the preservation or enhancement of the conservation area. On a challenge by the developers to this decision under section 245, it was argued that he had failed to have regard to a material consideration, namely the presumption in favour of development which does not cause demonstrable harm. This is now to be found in para 15, Planning Policy Guidance Note 1 — *General Policy and Principles* — issued by the Department of Environment, in the following terms:

> . . . There is always a presumption in favour of allowing applications for development, having regard to all material considerations, unless the development would cause demonstrable harm to interests of acknowledged importance. Except in the case of inappropriate development in the Green Belt the developer is not required to prove the case for the development he proposes to carry out; if the planning authority consider it necessary to refuse permission, the onus is on them to demonstrate clearly why the development cannot be permitted.

Mr Vandermeer said, [1990] 2 PLR 1, at p 6H:

> As a general proposition the Secretary of State and/or his inspector cannot, in my view, avoid the need to ask whether a particular proposal causes harm to an interest of acknowledged importance and to grant planning permission if it does not.

After quoting a passage from the judgment of Mr Lionel Read QC in *Steinberg* Mr Vandermeer said at p 7B:

> The need to address that duty — that is the duty under section 277(8) — does not however, in my judgment, relieve the Secretary of State and/or his inspector of the need to consider whether harm would be caused by the proposed development or, if so, whether the benefits of the proposed development outweigh that harm, nor do I read the judgment in *Steinberg* as so saying. Indeed, important though the provisions of section 277(8) are, they are only one of the material considerations that have to be taken into account and only require that "special attention" be paid to the matters set out in the subsection. . . .
>
> In addressing himself to the matters set in section 277(8) it is likely that an inspector would reach a view of whether harm had occurred. For instance, a finding that a development *enhances* the conservation area almost inevitably carries with it a finding that there would be no harm. If, on the other hand, there is a finding that a proposed development does *not preserve* the character

> or appearance of a conservation area it may well involve a finding that there is harm to the conservation area which would be harm to an interest of acknowledged importance. A

Mr Vandermeer then concluded that the inspector's decision letter made it clear that he did consider that the development would cause demonstrable harm etc, which was not outweighed by its benefits. The application was thus dismissed. B

In *Ward* v *Secretary of State for the Environment* (1989) 59 P&CR 486* the proposed development involved the loss of a private garden in a village which was a conservation area. On an appeal against the refusal of permission, the inspector decided to grant planning permission, apparently because the proposed development would, to an extent, enhance the conservation area. This court, however, concluded that his reasoning was not clear and quashed his decision. C

In the course of his judgment Woolf LJ said at p 492:

> . . . it seems to me that it should be borne in mind that the primary task which the inspector had to perform was to compare what is on the site at the present time with what is going to be on the site, assuming that the planning permission were to be implemented.

Major Crombie relies upon this observation in support of his submission, which I have already recorded, that the inspector has made a wrong comparison between the field and the garden as a site for the proposed building. D

Harrow London Borough Council v *Secretary of State for the Environment and Harrow School* (1989) 60 P&CR 525† was a decision of Sir Graham Eyre QC on December 12 1989. The proposed development involved the demolition of some buildings on a site, which overlapped the boundary between two conservation areas, and the construction of 21 houses and a theatre. Permission was refused and the developers appealed. The inspector who heard the appeal said that the main issue was whether the development would preserve or enhance the character or appearance of the conservation areas. He decided: E

(i) that the theatre would enhance the character and appearance of the conservation area; F

(ii) that the proposed housing would preserve the character of the conservation area; but

(iii) that the development would bring additional traffic to the area, which would be a disadvantage.

He balanced these competing factors and decided to allow the appeal.

Dismissing the application to quash that decision, Sir Graham Eyre at p 529 quoted section 277(8) and commented: G

> Clearly the inspector was exercising powers under the Act for the purposes of his determination of the appeals. As I understand the subsection, it does not set out the test nor does it support the proposition that in relation to a consideration of development proposals they must themselves preserve or enhance, or serve to preserve or enhance, the character or appearance of the conservation area. It imposes a duty to pay special attention to the desirability of preservation or enhancement. There may be cases where other important factors in the public interest are at work which render the development acceptable in a conservation area notwithstanding the conservation considerations. Such factors could be weighed by a decision maker provided he pays attention and special attention to the conservation objectives. H

*Editor's note: Also reported at [1990] 1 PLR 85.
†Editor's note: Also reported at [1990] 2 PLR 62.

A *South Lakeland District Council* v *Secretary of State for the Environment** another decision of Mr Lionel Read QC, given on February 26 1990, concerned an application for a new vicarage in a conservation area at Cartmel, Cumbria. The application for permission was refused and there was an appeal. The inspector said in terms that he was complying with the duty imposed on him by section 277(8). It was, however, argued before Mr Read that his decision was clearly not based on that test but on the "demonstrable B harm" test. The learned deputy judge held that the reasoning was not clear and quashed the decision.

He said:

> Plainly, in my judgment, such a duty (ie under section 277(8)) is different from and requires more than the duty under section 29 to have regard to the fact that a development is proposed in a conservation area as a material C consideration. Discharge of the former duty under section 277(8) must, in my judgment, also require the decision-maker dealing with such an application for planning permission to consider and decide whether the development would or would not preserve or enhance the character or appearance of the conservation area in question. Otherwise I do not see how he can discharge the duty of paying special attention to the desirability of preservation or enhancement.
>
> D An application is not however to be refused by reason of section 277(8) because it neither preserves nor enhances the conservation area's character or appearance. That is not the test proposed by the subsection. It may be refused for that reason or it may not. That is for the judgment of the decision-maker. It may be that, to give effect to the intention of Parliament, the decision-maker should give added weight to any conclusion that the development would not preserve or enhance the character or appearance of the conservation area, but that question does not fall for my decision. If, E however, the application is not refused for that reason, although it is desirable that the character or appearance should be preserved or enhanced, other reasons for not refusing it must, in my judgment, be given.

I have derived great assistance from the passages in these various judgments, and in general I agree with the sense of those I have quoted, though I do not necessarily agree with the precise wording of some of them. I can F summarise the conclusion I have reached about the proper approach which should be adopted to an application for planning permission for development in a conservation area as follows:

(i) If permission be sought for development on a site not within a conservation area, the person or body charged with deciding the matter — the local planning authority, the Secretary of State or his inspector on an appeal — is normally subject to one statutory duty, that imposed by section 29(1) G and, in the case of an appeal, section 36 of the 1971 Act. The departmental policy embodied in para 15 of PPG 1 is in such circumstances a perfectly proper approach to the carrying out of that statutory duty.

(ii) If, however, the site of proposed development is in a conservation area, then the decision-maker has two statutory duties to perform — those imposed by section 277(8) as well as section 29(1) of the Act. The question is, how are these duties to be performed in reconciliation with each other, and how do H they relate to the policy in para 15 of PPG 1?

(iii) In my opinion, in a conservation area the requirement under section 277(8) to pay "special attention" should be the first consideration for the decision-maker. It is true that the desirability of preserving or enhancing the character or appearance of the conservation area is, in formal terms, a "material consideration" within section 29(1). Since, however, it is a

*Editor's note: This decision was reversed by the Court of Appeal and is now reported at [1991] 2 PLR 97 *post*.

consideration to which special attention is to be paid as a matter of statutory A
duty, it must be regarded as having considerable importance and weight.

(iv) If, therefore, the decision-maker decides that the development will either enhance or preserve the character or appearance of the conservation area, this must be a major point in favour of allowing the development.

(v) Nevertheless, there will be some cases in which a development can
simultaneously enhance the character or appearance of a conservation area B
but nevertheless cause some detriment. That detrimental effect is a material consideration. The *Harrow* case, in which the inspector concluded that the quality of design of the proposed theatre would enhance the area, but that the additional traffic was a detriment, and that these two factors had to be weighed, the one against the other, is a good example of this. This seems to me to be a perfectly proper approach.

(vi) If, however, the decision-maker decides that the proposed development C
will neither preserve nor enhance the character or appearance of the conservation area, then it is almost inevitable that the development will have some detrimental, ie harmful, effect on that character or appearance. In lawyers' terms, the presumption derived from the policy in para 15 is then rebutted. As I have said, the conclusion that the development will neither enhance nor preserve will be a consideration of considerable importance and
weight. This does not necessarily mean that the application for permission D
must be refused, but it does, in my view, mean that the development should be permitted only if the decision-maker concludes that it carries some advantage or benefit which outweighs the failure to satisfy the section 277(8) test and such detriment as may inevitably follow from that.

Conclusion

If the approach I have indicated above is correct, the question then arises in E
relation to this case, is it apparent from the inspector's report that he and the Secretary of State have approached their decision on the appeal in the way I have suggested?

In his judgment, Hutchison J said that he agreed with Major Crombie's argument that "it is surprising that the inspector did not mention subsection (8) if in fact he had it in mind and was according proper weight to it". On the
other hand, the learned judge accepted the argument advanced by Mr Laws F
(then appearing for the Secretary of State) to the effect that since the major argument at the inquiry into the appeal centred on the effect of the conservation area, and since specific reference had been made to section 277(8), it was difficult to think that the inspector was not taking account of the duty imposed by that subsection. The judge referred to and quoted a number of passages in the inspector's report in which he summarised the evidence or
argument for the developer, for the local planning authority, and by Major G
Crombie as to the effect of the development on the conservation area.

Hutchison J concluded:

> If one turns then to the findings of fact and the conclusions, most of which I have quoted, there are references to the conservation area and a discussion of the merits or demerits of the proposed development in a context which, as it
> seems to me, inevitably leads to the inference that the inspector was alive to H
> and was giving proper weight to his section 277(8) duty.
>
> I see no reason to qualify this conclusion in the light of the three subsidiary matters specifically relied on, which form separate grounds, as to which I am persuaded that the arguments advanced by the respondents and summarised above are correct.

It was for these reasons that he rejected Major Crombie's argument relating to the section 277(8) duty.

A Despite the references to the conservation area throughout the inspector's report, I am unable to agree with Hutchison J that though the inspector made no express reference to the special attention he was required to pay by section 277(8), it is apparent from his report that he did fulfil his duty under that subsection. I cannot read into his report that he did approach the matter as I have suggested it should be approached.

I agree with Major Crombie's submission that the inspector fell into error in
B making a comparison between the effect of the proposed block of flats constructed on the field and that of a similar building constructed on the garden. For the reasons I have set out above in relation to this point, I agree with Major Crombie that the council's development brief would not have envisaged or permitted a building of anything like the bulk or scale of this block of flats on the garden. Thus, it follows that in making this comparison
C the inspector was posing an inappropriate question for himself.

I think the inspector may have been led into this error by his approach to the 1974 decision. Again, I agree with Major Crombie that it does seem that the inspector did not fully appreciate that the inspector in 1974 was reporting, and the Secretary of State was adopting, his recommendations on the basis that there was an existing planning permission in being. This aspect of the matter, however, seems to me to be peripheral to the main issue, to which I return.

D As I have said, I do not find it possible to infer from the inspector's report that he was paying "special attention to the desirability of preserving or enhancing" the character or appearance of the conservation area. But if this could be spelled out of his report, it is clear that there is no suggestion in the report that the building would enhance the character or appearance of the conservation area. Mr Howell argues that a conclusion that a proposed development will do no harm is equivalent to a conclusion that it will preserve.
E Even if that be correct (and adopting the approach of Mr Lionel Read QC, I doubt whether it is), this is not what, in my view, the inspector himself concluded. In para 42 of his report he said that the effect of the new building on the view looking southwards would "not be unacceptably great". This can only mean that he concluded that it would to some extent detract from that view. The inference to be drawn from that conclusion is that, if he was
F considering the section 277(8) question, he decided that the building would not preserve the appearance of the conservation area.

On this basis, if the inspector did comply with the section 277(8) duty, and formed the opinion that the building would neither enhance nor preserve the character or appearance of the conservation area, then he failed to carry out the necessary balancing exercise of weighing that important material consideration against such unexpressed benefits as the construction of the new
G building might bring.

For these reasons, I conclude that the reasoning in para 42 of the inspector's report was defective and did not satisfy the obligation imposed on him, and on the Secretary of State, by section 277(8). As I have said much earlier, the Secretary of State in his decision letter simply said that he "agrees with the inspector's conclusions and accepts his recommendation". The flaw in the inspector's reasoning, therefore, necessarily affects the Secretary of State's
H decision.

For these reasons, I would allow the appeal against Hutchison J's decision and quash the decision of the Secretary of State contained in his letter of November 25 1988.

STOCKER LJ: I entirely agree and add one short comment of my own, since we are differing from the conclusion of the learned judge.

In my view, the special attention which the inspector is bound to give to the

provisions of section 277(8) is of particular importance where the site A
concerned is of "such universal value that protecting it is the concern of all mankind". This fact, in my opinion, requires at least that it must be manifest from the terms of the inspector's report that very close consideration has been given to the provisions of this section, and it is insufficient that it is possible to spell out from the terms of the report that the inspector has, in some way, the terms of the section in mind.

Accordingly, I agree that the appeal should be allowed. B

SIR DENYS BUCKLEY: I agree both with the judgment delivered by Glidewell LJ and with the observation of Stocker LJ, which follows that judgment. I do so not without some regret, for para 42 of the inspector's report seems to indicate that the inspector was recommending what appeared to him to be a reasonable and realistic solution of the problem confronting
him, fairly balancing the amenity aspects and the development aspects of the C
matter, and I do not doubt that he had the terms of section 277(8) well in mind. I feel, however, the force of Glidewell LJ's criticism of that paragraph of the report near the end of his judgment.

I also would allow this appeal.

Appeal allowed with costs.

D

North Wiltshire District Council *v* Secretary of State for the Environment and others E

QUEEN'S BENCH DIVISION
Mr Lionel Read QC (sitting as a deputy judge of the Queen's Bench Division)
February 26 1991

**Material considerations — Development plan — Emerging local plan —
Whether inspector wrongly preferred policy of draft local plan to more F
restrictive policy in adopted plan — Previous contrary decision by another inspector — Inspector not giving reasons for reaching different conclusion — Decision on that ground to be quashed**

The second respondents, Mr and Mrs Clover, appealed against the refusal by North Wiltshire District Council of their application for planning permission
for the erection of a dwelling and garage on land within the walled garden of G
Notton Lodge, a Grade II listed building in the village of Notton, near Chippenham, Wiltshire. The reasons for refusal included that the proposal was contrary to policy H14 of the Western Wiltshire Structure Plan 1981 and policy H8 of the draft North Wiltshire Local Plan. Policy H14 of the structure plan provided that "only very limited residential development within the physical limits of the village will normally be permitted. Such development
will be acceptable only if it is in scale and harmony with the character of the H
settlement and provided there is no adverse effect on the local environment." Policy H8 of the draft local plan was in similar terms. At the date of the appeal to the Secretary of State for the Environment the draft local plan had not yet been placed on deposit; however, it was agreed to be a material consideration for the purposes of section 29 of the Town and Country Planning Act 1971 (section 70 of the Town and Country Planning Act 1990). The relevant policy of the adopted local plan, the Chippenham Local Plan 1987, was more

A restrictive and did not allow for even such limited residential development in the village of Notton. The appeal, which was dealt with by way of written representations, was allowed by the inspector who concluded, in effect, that the appeal site was within the physical limits of Notton and that the proposed further dwelling was acceptable.

The local planning authority applied to the High Court under section 245 of the 1971 Act (section 288 of the 1990 Act) to quash the inspector's decision. It
B was submitted that the decision was wrong in law for four reasons: (1) the inspector failed to pay regard to the provisions of the development plan; (2) he proceeded on a false basis as to the approach of the council to the local plan; (3) he failed to pay regard to a material consideration, namely a decision in 1982 by another inspector who, in considering the same structure plan policy, had refused planning permission for the erection of a dwelling in the walled garden at Notton Lodge; and (4) he had failed to give reasons as to his
C approach to the 1982 decision.

Held The application was granted.

1. In view of the written representations the inspector was entitled to find as a fact that the council regarded the draft local plan policies as superseding, for development control purposes, the more restrictive policies in the adopted local plan: see pp 72A-73A.

2. The inspector did not fail to have regard to the adopted local plan: he
D expressly referred to it and its more restrictive policies, but he was entitled, as he did, to attach greater importance to the structure plan and the draft local plan. That was a matter of weight for him and he had not erred in law: see p 73B-E.

3. However, the inspector had failed to mention the 1982 decision, nor was there anything in his decision letter from which it could necessarily be inferred that he had considered it. The 1982 decision was directly in point, as the
E previous inspector, who had given reasons for his decision, had concluded that the site of that appeal lay outside the physical limits of Notton. The result was that the council were left with two diametrically opposite decisions on appeal without any explanation of which they should, in reason and in justice to other applicants, follow. The inspector, as judge of fact, was entitled to reach a different conclusion from his predecessor without erring in law or without necessarily being unreasonable, but his reasons for so doing ought to have been given or be otherwise apparent: see pp 74E-75E. The inspector had
F not adequately dealt with a substantial point raised by the council and on that one ground his decision would be quashed.

Case referred to in the judgment

Seddon Properties Ltd v *Secretary of State for the Environment* (1978) 42 P&CR 26; 248 EG 951; [1978] JPL 835

G **Application under section 245 of the Town and Country Planning Act 1971**

This was an application by North Wiltshire District Council under section 245 of the Town and Country Planning Act 1971 (section 288 of the Town and Country Planning Act 1990) to quash a decision by the first respondent's inspector who had allowed an appeal by the second respondents, Mr and Mrs Keith Clover, against a refusal by the council to grant planning permission for the erection of a dwelling and garage on land at Notton, Chippenham, Wiltshire.

H *Timothy Straker (instructed by Sharpe Pritchard, agents for the solicitor to North Wiltshire District Council) appeared for the applicants.*

John Laws (instructed by the Treasury Solicitor) appeared for the first respondent, the Secretary of State for the Environment.

The second respondents, Mr and Mrs Keith Clover, did not appear and were not represented.

The following judgment was delivered.

MR LIONEL READ QC: This is an application by North Wiltshire District A
Council, who are the local planning authority for their area, under section 245
of the Town and Country Planning Act 1971 [section 288 of the Town and
Country Planning Act 1990] to quash a decision by an inspector appointed by
the Secretary of State for the Environment (the first respondent).

By that decision, given by letter dated June 5 1990, the inspector allowed an
appeal by Mr and Mrs Keith Simon Clover (the second respondents) against a
refusal by the council to grant planning permission for the erection of a B
dwelling and garage on land at Notton, Chippenham, and granted planning
permission as sought. The appeal was dealt with by written representations.
The second respondents are not represented on this application.

Notton is, as the inspector described it, a loosely-knit settlement. The
appeal site lies within the walled garden of a listed building called Notton
Lodge. The creation of an access to the highway requires the demolition of a C
short length of garden wall. The outbuilding of the main house had recently
been converted to a dwelling known as "The Coach House".

Planning permission was refused for five reasons, the fifth of which, being
concerned with highway matters, is not relevant to this appeal, and I read only
the first four reasons:

> 1. The proposal, by reason of its location outside what can be reasonably
> defined as the physical limits of the village and its effect on the character of D
> the settlement and adjacent development is contrary to Policy H14 of the
> Western Wiltshire Structure Plan and Policy H8 of the Draft North Wiltshire
> Local Plan 1988 which seek to restrict new development to sites within the
> physical limits of the village so as to preserve its character and prevent
> undesirable and unnecessary expansion into open countryside.
>
> 2. The site is in a countryside location where, in the interests of the rural
> amenity of the area, it is the policy of the local planning authority to restrict E
> residential development to that which is justified in connection with the
> essential needs of agriculture or forestry.
>
> 3. In the opinion of the local planning authority the proposal would form
> a significant and undesirable encroachment of residential development upon
> an attractive and historic walled garden within a predominantly rural area,
> which would be detrimental to the character of this area in particular, and
> rural amenity in general, and create a strong precedent for erecting further
> dwellings on adjoining land and similar unsatisfactory development outside F
> the physical limits of the village.
>
> 4. The proposal, by virtue of its siting, proposed means of access and
> visibility splays and effect on the attractive and interesting walled garden
> within the historic curtilage of Notton Lodge would detract from the
> character and appearance of the wall (Grade II listed), adversely affect the
> setting of Notton Lodge and outbuildings and be contrary to Policy B3 and
> B4 of the Draft North Wiltshire Local Plan 1988 which seeks to preserve G
> buildings of architectural or historic interest and only permit development
> which does not adversely affect their character or the general environment of
> the area.

The written representations comprise the following documents in
particular: the notice of refusal and notice of appeal; the planning officer's
report to the committee; the council's, as they called them, "concluding H
submissions and comment" dated January 22 1990 (which were sent to the
Department); a written statement by a planning consultant acting for the
applicants for planning permission dated March 19 1990 and additional
comments by the council in a letter to the Department dated April 3 1990.

Mr Straker, who appears for the council, submits that the inspector's
decision was wrong in law for four reasons: (1) he failed to pay regard to the
provisions of the development plan; (2) he proceeded on a false basis as to the

A approach of the council to the local plan; (3) he failed to pay regard to a material consideration, namely a previous relevant decision by another inspector; and (4) he failed to give reasons as to his approach to that decision.

The council's case before this court centred upon the provisions of three plans: the structure plan, the adopted local plan and a draft emerging local plan. It is necessary that I should briefly identify those plans and their relevant provisions.

B The appeal site lies in an area which is covered by the Western Wiltshire Structure Plan, which was approved by the Secretary of State in 1981. The relevant provisions of that plan, which is part of the development plan within the meaning and for the purpose of the Town and Country Planning Act 1971, are policies H13, H14 and H15. The effect and relevant wording of those policies may be briefly summarised. Policy H13 allows development of small C groups of dwellings in villages which satisfy certain criteria. Notton is not one of those villages. In relation to villages which do not meet those criteria policy H14 provides that:

> only very limited residential development within the physical limits of the village will normally be permitted. Such development will be acceptable only if it is in scale and harmony with the character of the settlement and provided there is no adverse effect on the local environment.

D In a logical sequence of provisions, policy H15 then deals with new dwellings in the countryside

> . . . unrelated to any established village [which] will normally not be permitted unless justified in connection with the needs of agriculture or forestry.

E Policy H14, therefore, applies to the village of Notton. The issues raised by this policy are whether it applies to the appeal site as being within the physical limits of the village and, if so, whether the development proposed would be acceptable by reason of consideration of scale, harmony and effect on the local environment.

The appeal site also lies in an area covered by the Chippenham Local Plan, which the council adopted in 1987 and which is therefore also part of the F development plan. It was not in dispute in the representations made by the parties to the inspector, or in the submissions to this court, that the policies of the adopted local plan are more restrictive than those of the structure plan.

Policy H5 of the adopted local plan allows housing development in certain identified villages subject to considerations of adverse effect and parking and highway matters; but Notton is not one such village. There is no policy which allows for very limited residential development in other villages.

G The next policy, H6, refers to new dwellings in the countryside, which would be justified only in connection with the essential needs of agriculture and forestry. Hence, on the basis of the policies in the adopted local plan, the development under appeal would not be in accordance with policy and would fall to be determined as an exception to policy.

At the time of the appeal to the Secretary of State the council were in the H course of preparing for adoption another local plan, covering the area of the appeal site, called the North Wiltshire Local Plan. A draft had been prepared for consultation, and modification to it had been made following consultation. It had not, however, been put on deposit for the purposes of objection. Accordingly, it had not been adopted and had not yet become part of the development plan in succession to the Chippenham Local Plan; it was, that is to say in language used by those versed in planning matters, emerging. As such it was, as was agreed by the parties to the appeal to the Secretary of

State, a material consideration for the purposes of section 29 of the Town and Country Planning Act 1971 [section 70 of the 1990 Act]. A

The policies of the draft local plan closely follow the provisions of the structure plan. Policy H7 allows residential development within the built-up framework of certain identified settlements. Notton is not so identified. In relation to those settlements which are not so identified, including therefore Notton, policy H8 states that:

B

> only very limited residential development within the physical limits of the village will normally be permitted,

subject thereafter, in short, to considerations of scale, harmony, character and adverse effect on the local environment. In as much, therefore, as the draft local plan was a material consideration on the appeal to the inspector, the same issues arose from its provisions as those which arose from the structure plan. C

The inspector referred to these three plans in para 2 of his decision letter in these terms:

> Planning policies relevant to this proposal are found within the West Wiltshire Structure Plan which was approved in 1981 and in the emerging North Wiltshire Local Plan which the Council appear to regard as superseding, for development control purposes, the more restrictive policies in the Chippenham Local Plan which they approved in 1987. The representations include extracts from all these plans and it seems to me that policy H14 of the Structure Plan and H8 of the draft Local Plan are of particular importance. D

The critical words in that passage, for the purpose of Mr Straker's principal argument, are: "which the Council appear to regard as superseding . . . the Chippenham Local Plan". E

The inspector does not in terms again refer to the policies of these plans. Nevertheless, at para 3 of his letter he framed what he called the main issues in the case as follows:

> . . . whether or not, taking account of these policies, the proposed house would amount to intrusive development harming either the rural amenity of the local scene or the setting of Notton Lodge which with its former outbuildings and boundary walls is a listed building. F

He then concludes his determination on the first part of this main issue at the end of para 4 of his letter with the following passage:

> No boundary has been defined for the loosely-knit settlement of Notton, but having approached it from various directions I am drawn to the conclusion that it would be unrealistic not to regard the former garden, the group of associated historic buildings and the more recent dwelling of somewhat suburban design immediately to the south as one part of a village to whose character fields penetrating its core are of great importance. Accordingly, though undoubtedly not infill in the usual sense of the word, it is my opinion that the addition of a further dwelling within this group need not in principle conflict with the Council's policies. I am in no doubt that the proposal cannot be regarded as the sort of sporadic or haphazard development in open countryside which those policies very properly aim to prevent. G H

Mr Straker mounts his challenges to the inspector's decision on those passages. The first of his challenges has two limbs and arises under the first two of his principal arguments, which I have summarised. He draws attention to the inspector's words in para 2 of his decision letter, which I have already highlighted, and submits, first, that the inspector was wrong in what he said

A and, second, that even if he were correct he could not ignore the adopted local
plan unless it was not material to his determination of the application.

The thrust of the council's case on this first submission by Mr Straker is that
the inspector was erroneous, as the grounds of the application to this court put
it, in proceeding on the basis that the council regarded the adopted local plan
policies as superseded by the policies in the draft local plan. The foundation
B for that argument lies ultimately in the point made by the council in their
additional comments.

The agents for the applicants for planning permission had dwelt at some
length in their representations on the policies of the structure plan, the
adopted local plan and the draft local plan. They expressed their
understanding that the council had accepted that the draft local plan provided
the more appropriate interpretation of structure plan policies at a local plan
C level than the adopted local plan. They referred to an appeal decision in 1989,
which was appended to these representations and concerned a site in the village
of Notton. The decision letter in that appeal, so far as is relevant, reads:

> It was agreed by both parties at the hearing that the relevant policies were H14
> of the Western Wiltshire Structure Plan and H8 of the draft North Wiltshire
> Local Plan. The wording in both was similar, and reflected a flexible
> approach to limited housing within the limits of smaller villages, in line with
> D government advice and superseding a more restrictive policy in the adopted
> Chippenham Local Plan.

These consultants concluded with the observation that they did not propose to
consider further the policies and provisions of the adopted local plan.

The council responded in their additional comments with the observation:

> It is a matter of fact that the Chippenham Local Plan is a material document
> E and cannot be discounted. The weight to be attached to it is a matter for the
> inspector to consider in the consideration of the merits of this appeal.

As a statement of fact and law this comment cannot be faulted. It helped the
inspector, however, not at all on the question whether the council accepted
that the local policies which should guide determination of the appeal were
those in the draft local plan rather than those in the adopted plan. A question
F and answer at an inquiry might have resolved that potential issue. The
inspector had to decide it on the written representations before him. Those
representations included the planning officer's report to his committee where
he said:

> In planning policy terms, Notton does not have a defined built-up framework
> in the draft North Wiltshire Local Plan but Policy H8 refers, like the Western
> Wiltshire Structure Plan, to only very limited residential development within
> G the physical limits of a village, which amends a more restrictive policy
> contained in the adopted Chippenham Local Plan. The physical limits of any
> village in these circumstances must be decided on its merits.

No other reference is made in this report to local plan policies and, it is to be
observed, the only reference to local plan policies in the council's reasons for
refusal is to policies of the draft local plan. In their concluding submissions
H and comment, the council drew attention to the fact that in the context of the
adopted local plan Notton is not a village listed in policy H5 and, as such, is
regarded as countryside under policy H6. Nevertheless, the council made the
further point in those submissions that:

> . . . for the purposes of the application Notton is considered a village in the
> context of policy H8 of the Draft North Wiltshire Local plan.

Those in short were the facts before the inspector. He concluded on them

that the council appeared to regard the draft local plan policies as superseding, A
for development control purposes, the more restrictive policies in the adopted local plan. That was a finding of fact. There was, in my judgment, evidence before the inspector in the written representations upon which he could reasonably have made that finding. He did not accordingly err in law in reaching that conclusion.

I turn to the second limb of Mr Straker's argument, which related to the
adopted local plan. Mr Straker submits, and it cannot be in doubt, that section B
29(1) of the Town and Country Planning Act 1971 obliged the inspector to have regard to the provisions of the Chippenham Local Plan, which was part of the development plan, so far as they were material to the application, as some of them plainly were. Nevertheless, it is trite law that, provided that he had regard to those provisions, the weight to be given to them was a matter for him and he was not obliged to follow them provided that he had and gave a C
sustainable reason for not doing so.

In my judgment, the inspector did have regard to the adopted local plan policies. He expressly referred to that plan and to its more restrictive policies, but his decision was not determined by them. In that conclusion he was plainly influenced by the fact that the council appeared to him to regard them as superseded by the draft local plan policies for development control purposes.
It is implicit in his decision that he saw no reason to disagree with the council. D

I agree with the point shortly put by Mr John Laws, for the Secretary of State, that the engine of the council's submissions to the inspector, by their written representations, was the structure plan and the draft local plan. The policies of the adopted local plan were different from and more restrictive than those of these two plans and particularly of the structure plan. It is clear that the inspector's decision was guided by the policies of these two plans,
which he said were of particular importance. In that approach he did not fail to E
have regard to the adopted local plan. He attached greater importance to the different provisions of the structure plan and the draft local plan. That was a matter of weight for him, and I find no error of law in that decision.

The second main group of Mr Straker's submissions related to the way in which the inspector dealt with, or more strictly did not deal with, the
provisions of policy H14 of the structure plan and a previous decision of F
another inspector given in 1982. To summarise Mr Straker's point, if I may presume to compress his succinct and lucid argument, he submits that the inspector did not address the issue raised by the structure plan and, incidentally, by the draft local plan, namely whether the appeal site was within the physical limits of Notton, and either ignored a previous decision of another inspector who had effectively addressed that issue and decided that
the appeal site was not within those limits, or failed to give any reason why he G
disagreed with that decision. While those arguments challenged the inspector's decision from different directions, they are closely related and hence I consider them together.

The earlier decision referred to is dated February 4 1982. It is not the subject of any comment by the council in their representations and appears in those
representations only as an item of planning history. That decision related to an H
application for planning permission to erect a dwelling in the walled garden of Notton Lodge. The appeal site in the case before the court comprises the southern part of that walled garden.

At the time of that decision there was no local plan for the area of Notton and the inspector in that appeal was solely concerned with structure plan policy. He regarded the main issues in that appeal as being — I read from para 2 of his decision letter:

> . . . whether or not the proposed development can be regarded as infilling within the physical limits of an existing settlement and, if not, whether or not there is adequate justification for permitting the development as an exception to the normal requirements of the Western Wiltshire Structure Plan . . .

On that issue he concluded at para 5 of his letter that the site of that appeal lay outside the physical limits of Notton and the development could not be regarded as infilling. In his view, in para 6 of his letter, there were

> very strong planning objections to a proposal which would consolidate existing sporadic development and erode the open rural character of the locality contrary to the policies of the . . . structure plan.

In the appeal with which the court is concerned, the inspector's conclusion is to be found at the passage in para 4 of his letter, which I read earlier in this judgment. Hence, he said at para 5 that "the development would not have a harmful impact upon the countryside". It is the fact that the inspector in the appeal under challenge did not in terms say that the appeal site was, or was not, within the physical limits of Notton. I have no doubt, however, that he had well in mind policy H14 of the structure plan and policy H8 of the draft local plan. He concluded that the former garden of Notton Lodge, which encompasses the appeal site, was, with the group of associated historic buildings and the more recent dwelling, "part of a village".

In my judgment, this is a sufficient conclusion that it lay within the physical limits of Notton. Mr Straker draws attention to the inspector's use of the word "boundary" and submits that this is not the same thing as physical limits. But with due respect to Mr Straker's submission, that, in my judgment, is a semantic point. If the appeal site is, as the inspector found, part of the village, I see no distinction between that finding and an express finding that it was within the physical limits of the village.

The ultimate, and I think most difficult, question concerns the 1982 decision. The council clearly raised this decision for the inspector's consideration. It was directly in point. It raised, in my judgment, a substantial issue. The question is whether that issue was sufficiently dealt with. The inspector did not mention that decision, nor is there anything in his decision from which it may necessarily be inferred that he considered it. I am bound to observe that that is at least regrettable. Since the 1982 decision was directly in point, and the inspector reached a conclusion which was, on any view, quite contrary to that of his colleague on the earlier appeal, an overt recognition of those facts would at least have satisfied the council that the issue which they had raised had been addressed. Nevertheless, the question remains whether there has been an error in law and, if so, whether the decision under challenge should for that reason be quashed.

Mr Laws submits that the question is whether the parties to the appeal know from the decision letter why the inspector reached his decision, not whether they know why he has reached a conclusion different from that of the 1982 inspector. He submits that this inspector's reasons for his conclusions are clear and sufficient as a matter of law and that, in so far as the reasons for those conclusions are clear but differ from those of his predecessor on the 1982 appeal, all that he could effectively have added were words to the effect that he differed from his predecessor for those reasons. Those additional words would, however, add nothing of substance to what the inspector did in fact say.

Mr Straker submits, however, that the inspector could have given a number of reasons for differing from his predecessor but did not do so. Those reasons arise, in short, from the way in which the 1982 inspector dealt with the issues at para 4 of his decision letter or, conceivably perhaps, from the subsequent conversion of The Coach House. The thrust of Mr Straker's submission is that if this inspector had reasons for differing from his predecessor he does not give them, and accordingly they are not known to the council. A

In weighing these submissions, I remind myself of the classic exposition of the relevant test given by the late Forbes J in *Seddon Properties Ltd* v *Secretary of State for the Environment* (1978) 42 P&CR 26, at p 27, where the learned judge said that the reasons which the Secretary of State must give for his decision, and this necessarily applies equally to his inspector exercising transferred powers, must be proper and adequate, clear and intelligible "and deal with the substantial points that have been raised". B

In my judgment, the inspector in this case has failed to comply with that requirement. I do not think it is enough for him to reach and express a conclusion which is different from that of his predecessor on essentially the same, if not identical, facts without any overt or necessary recognition that he has addressed that previous decision and without some comprehensible explanation of why he disagrees with it. C

The reality of the matter is that the council are left with two diametrically opposite decisions on appeal without any explanation of which they should, in reason and in justice to other applicants for planning permission, follow. I do not agree that in the face of the 1982 decision, on apparently identical facts, it was enough for the inspector in the appeal under challenge to give reasons for his decision without in any way addressing the reasoning of his predecessor. That does not adequately deal with the substantial issue raised by the council in the form of a decision directly in point which supports their contention. D

The matters with which the two inspectors had to deal were ones of facts which required judgment. As a judge of fact the inspector in the case under challenge was entitled to reach a different conclusion without error in law or without, in particular, necessarily being unreasonable; but reasons for the difference ought to be given or be otherwise apparent. In my judgment, the reasons for this inspector's difference from his predecessor are not apparent and they certainly have not been given. E

I have considered whether any useful purpose is to be served by remitting this matter back for redetermination. In my judgment, there is. I do not agree that it would be enough for this inspector to say expressly that which may be necessarily implicit in his decision; that is to say, simply that he disagrees with his predecessor for the reasons he has already given. The 1982 inspector gave reasons for his decision. I think it unlikely that the inspector in the present case could deal adequately with that decision without addressing those reasons and saying whether and why he disagreed with them or whether he had some other, and if so what other, reasons for reaching a contrary conclusion. As his decision at present stands, those are matters of speculation. For that reason above all, it is in my judgment deficient in proper, adequate, clear and intelligible reasons for dealing with a substantial point which the council had raised. F G

For those reasons, and on that one ground, I hold that the inspector erred in law and that his decision should be quashed. H

Application granted with costs.

Uttlesford District Council *v* Secretary of State for the Environment and another

QUEEN'S BENCH DIVISION

Mr Lionel Read QC (sitting as a deputy judge of the Queen's Bench Division)

March 4 1991

A **Dwellinghouse — Erection of garage building — Use of part of garage building as annex to house — Whether breach of planning control — Whether creation of separate planning unit — Appeal by council dismissed**

The second respondent, Mr Rowland White, is the owner of land and a house at Woodview, Cornells Lane, Widdington, Essex. Planning permission was granted in October 1988 for the erection of a house with a detached garage. B The garage was erected to a height some 0.6 m above that approved and following a second application planning permission was granted in April 1989 for the retention of the building on condition that it was used for domestic purposes only incidental to the enjoyment of the dwelling and not for a residential annex or a separate dwelling. At ground level the garage building is now divided into a conventional garage and, alongside it, a kitchen area and breakfast bar leading to a staircase. On the first floor there is a bathroom and C small bedroom. The second respondent applied for planning permission for the change of use of the garage to provide living accommodation in conjunction with the existing dwelling. In October 1989 the applicant council issued an enforcement notice alleging breach of planning control.

By his inspector, the Secretary of State for the Environment allowed appeals by the second respondent against an enforcement notice issued by the appellant council and against a refusal of planning permission; the enforcement notice, which was quashed, had required the removal of the D domestic features in the garage. The inspector had decided that the application was not strictly necessary as no material change of use is involved if a domestic garage within a residential curtilage is used for living accommodation in connection with a dwelling. The inspector treated the appeal against the refusal of planning permission as if an application for planning permission had been made under section 63 of the Town and Country Planning Act 1990 for the retention of the garage at its increased E height without compliance with the condition attached by the April 1989 condition. The council applied under section 288 of the 1990 Act to quash the first respondent's decision allowing the appeal against the refusal of planning permission on the grounds that the inspector had erred in treating the appeal in the manner he had and that he was wrong in law.

Held The application was dismissed.

F In treating the appeal as if an application had been made to retain the garage without compliance with the planning condition, the inspector did not have to decide and did not decide whether planning permission should be granted for the use proposed in the application submitted to the council: see p 79E. The inspector addressed and decided the first issue as to whether the application involved a material change of use.

The identification of a separate planning unit is a question of fact to be decided on the evidence: see p 82F. The annex in the garage building did not G constitute a separate planning unit, the proper test being that of a single family occupation. Having decided that the application did not involve a material change of use, it was unnecessary for the inspector to consider the second stage whether the user was incidental to the dwellinghouse: see p 83.

In relation to the planning condition, the inspector was entitled to find that it was contradictory with the planning permission to which it was attached: see p 83. Although the inspector came to the right conclusion for the wrong reason, in concluding there had been no material change of use of the garage building, it was not necessary to quash the decision. A

Cases referred to in the judgment

Burdle v *Secretary of State for the Environment* [1972] 1 WLR 1207; [1972] 3 All ER 240; (1972) 70 LGR 511; 24 P&CR 174; [1972] EGD 678; 223 EG 1597, DC
Emin v *Secretary of State for the Environment* [1989] JPL 909
Wallington v *Secretary of State for Wales* [1991] 1 PLR 87, CA B

Application under section 288 of the Town and Country Planning Act 1990

This was an application under section 288 of the Town and Country Planning Act 1990 to quash a decision of the first respondent, the Secretary of State for the Environment, who by his inspector had allowed an appeal against a refusal of planning permission by the applicant council following an application made by the second respondent, Mr Rowland White.

Peter Village (instructed by Jameson & Hill, of Hertford) appeared for the applicants, Uttlesford District Council. C
Guy Sankey (instructed by the Treasury Solicitor) appeared for the first respondent, the Secretary of State for the Environment.
The second respondent, Mr Rowland White, appeared in person.

The following judgment was delivered.

MR LIONEL READ QC: This is an application under section 288 of the Town and Country Planning Act 1990 by Uttlesford District Council, who are the local planning authority for their area, to quash a decision by one of the inspectors of the Secretary of State for the Environment, the first respondent, given on an appeal by Mr Rowland White, the second respondent, against a refusal of planning permission by the council. The second respondent, who appears in person on this application, converted the garage of his house to provide some very small separate living accommodation for his mother, but, unfortunately, she died before the conversion was completed and the intention, as I understand it, is that it should now be occupied by his mother-in-law. It is in short, as the second respondent aptly described it in his application for planning permission, a "granny annex". D E

It is a sad commentary that such a modest proposal for a family purpose should end up in the High Court, but Mr Village, who appears for the council, has said that the council are concerned with the principle involved in the inspector's decision, if there be a principle, and its implications for proposals of a similar kind in their district. The relevant facts are a little complex, but, stripped of matter not relevant to the application to this court, they may be shortly summarised. F

Planning permission was granted in October 1988 for the erection of the house in which the second respondent now lives with a detached garage. This garage was built to a height greater than that which was permitted and planning permission for the retention of the garage with an increased height was granted in April 1989 subject to the condition that the garage should be used for domestic purposes only incidental to the enjoyment of the dwelling and not for any separate industrial commercial business use or as a residential annex or separate dwelling. In the event the second respondent converted the garage so as to preserve some garage space, but also to provide at ground-floor level and above what had been intended as storage space some small separate living accommodation. G H

In March 1989 the second respondent applied for planning permission for this conversion. The application was expressed to be for "change of use for detached garage to private living accommodation to be used in conjunction with existing dwellings (in short 'a granny annex')". That application was

A refused by the council in May 1989 and the second respondent appealed against the refusal in August 1989. In October 1989 the council issued an enforcement notice in respect of the conversion, which the second respondent also appealed under para (*a*) of section 88(2) of the Town and Country Planning Act 1971 (section 174 of the 1990 Act), that is to say, seeking planning permission for the development enforced against as well as under
B para (*h*). The inspector's decision on that appeal to quash the enforcement notice is not the subject of the council's application to this court and it is unnecessary for me to refer to it again.

Both appeals were dealt with by written representations. Following the inspector's considerations of those representations an officer of the Department of the Environment wrote to the parties on May 14 1990 conveying, somewhat unusually but I believe helpfully, the inspector's initial
C reactions for the parties' comments. Among his observations reported in that letter were the following:

> 3. He

— that is to say, the inspector —

> considers that the use of the building as a residential annexe to the existing dwelling would involve no material change of use of the dwelling and its
> D curtilage and does not constitute development for which planning permission is required because the use of the planning unit was, and remains, residential. If

— which is not the present case —

> the garage were to be severed from the house or to be independently occupied then, of course, this position would change. It follows that the application
> E subject to the s 36 appeal was unnecessary.
>
> 4. Although no development would be involved the question remains whether use in the manner proposed would breach the condition attached to the approval of application UTT/0107/89

— which I interpolate to explain is the one which permitted the increased height of the garage —

> F In this connection the Inspector commented that . . .
>
> B. the condition as drafted is arguably contradictory since use as a residential annexe (which it seeks to prohibit) would be a use incidental to the use of the dwelling (which it permits).

This letter is relevant to the extent that the inspector's subsequent decision is to be read in the light of, and reflects, these initial comments. I have omitted the additional criticism of the condition given in para 4A of this letter, since it
G is of no direct relevance to the issues which I have to determine. But I need to refer to para 8, which reads:

> If, the council, in light of the observations at 4A and 4B above, intend to assert the continued validity of the condition then it seems to the Inspector that a proper course for him to adopt would be to treat the application as being made under s 31A of the amended 1971 Act — for the development of land without complying with a condition subject to which a previous
> H permission has been granted. That would enable the validity of the condition as well as the planning merits of [its] removal to be considered.

The council responded to these initial comments by a letter of July 5 1990 from their director of planning. He expressed his agreement with the view of the inspector given at para 3 of the department's letter that "the use of the building as a residential annexe to the existing dwelling would involve no material change of use . . . and does not constitute development for which planning permission is required". He also agreed with para 4B of the department's letter and continued:

> It is the use of the unit as a separate dwelling which the Council seeks to prohibit. Planning permission has been granted for a garage — if it is sold away separately from the main house then it will no longer be within its curtilage, so any residential occupation can then not be ancillary. In this instance planning permission would be required for a conversion to dwelling and any application would be considered on its merits.

He concluded by asserting the validity of the condition subject to which the April 1989 permission was granted.

The inspector then gave his decision by letter dated August 29 1990. By that time the Town and Country Planning Act 1990, which consolidated the Town and Country Planning Act 1971 and other enactments, had come into force. As I have said, he quashed the enforcement notice. On the appeal against the refusal of planning permission he decided that he should treat the application for planning permission as if it were made under section 63 of the 1990 Act for the retention of the building at its increased height without compliance with the condition attached to the April 1989 consent. He considered this was the appropriate course rather than under section 73 of the 1990 Act (previously section 31A of the 1971 Act), as suggested in the department's letter of May 14 1990, since the consent on which the condition was imposed had been implemented. The council make no criticism of this course of action. Indeed it was, as Mr Village for the council accepts, agreed by the parties.

In the result, by treating the application in this way the inspector did not have to decide and did not decide whether planning permission should be granted for the use proposed in the application submitted to the council. On the, as it were, amended application the only question was whether the condition attached to the April 1989 permission for the retention of the garage should be removed or modified. On their second principal head of argument the council challenged the correctness in law of the way in which the inspector decided that question. I deal with that argument later in this judgment, but at this stage I am concerned to identify and then address the council's first principal ground for challenging the inspector's decision.

This first principal ground of challenge is directed to the inspector's reasons for determining the application under appeal in the way that he did. That reason was foreshadowed in para 3 of the department's letter of May 14 1990. It was formally incorporated as part of his decision at para 9 of his decision letter. It is this paragraph which is the subject of challenge. That paragraph reads as follows:

> The application the subject of this appeal was not strictly necessary in my view since no material change of use is involved if a domestic garage within a residential curtilage is used for living accommodation in conjunction with the dwelling — a point made in paragraph 3 of my earlier letter and — again — one from which the Council do not dissent.

In arguing that this paragraph reveals error of law Mr Village conceded at the outset of his argument that the council not only had not dissented from the proposition there stated by the inspector but had agreed to it. This is apparent from the passage in the council's letter of July 5 1990, which I have already read. Hence the council agreed on the argument now addressed to the court to an error of law by the inspector, but, Mr Village argues, an error of law cannot be agreed by the parties to an appeal. Now contending that there is such an error of law, the council by this application accordingly bring the matter before the court for correction.

I agree with this part of Mr Village's argument to this extent. If there be

A error in law by the inspector such that his decision was outside the powers given to him by statute, the council cannot bring his decision within those powers by any agreement to what the inspector in fact did. The burden of Mr Village's submissions is that the inspector erred in law because he did not approach correctly the question which he had to determine on the application under appeal, reached a wrong conclusion on that question and failed to give adequate reasons for it. Those submissions were pervaded throughout by Mr
B Village's contention that there is no rule of law that, as para 9 of the decision reads, "no material change of use is involved if a domestic garage within a residential curtilage is used for living accommodation in conjunction with the dwelling". That, he submits, is not only an over-simplification; it is a manifestly wrong approach.

In deploying his argument on those submissions Mr Village begins with the
C definition of development, now to be found in section 55 of the 1990 Act. Subsection (1) defines development as meaning, so far as here relevant, "the making of any material change in the use of any buildings or other land". Subsection (2) then provides that the operations or uses of land therein set out "shall not be taken for the purposes of this Act to involve development". The use at para (d) is here relevant, that is to say, "the use of any buildings or other land within the curtilage of a dwellinghouse for any purpose incidental to the
D enjoyment of the dwellinghouse as such".

Mr Village submits that section 55 required the inspector to follow a two-stage approach by which he should first have considered whether the use of the garage as proposed constituted a material change of use and hence, at that stage, development, and, if it did, whether, as a second stage, that use was for a purpose incidental to the enjoyment of the principal dwellinghouse as such. In support of his submission Mr Village referred to the judgment of
E Farquharson LJ in *Wallington* v *Secretary of State for Wales* given on November 7 1990 and so far unreported*. At p 3 of the transcript of his judgment Farquharson LJ said†:

> The section contemplates a two-stage approach, first, has there been any material change in the use of, in this case, the cottage? If not, then the enforcement notice is bad and the appellant would be entitled to succeed. If
> F there has been a material change of user, there will still not be a development within the meaning of subsection (1) if that user is incidental to the enjoyment of the dwellinghouse as such. In the argument before us the questions posed by these two subsections have been taken together, but they should be dealt with separately.

Slade LJ at pp 6 and 7 of the transcript§ similarly stressed the need for the inspector to adopt this two-stage approach, though he referred to the two
G stages as "two separate questions".

Mr Village then submits that, if this two-stage approach had been followed, which, he submits, it was not, the inspector should first have determined the correct planning unit. On that question Mr Village cites the judgment of Bridge J (as he then was) in *Burdle* v *Secretary of State for the Environment* [1972] 1 WLR 1207 at p 1212. There the learned judge sketches out three broad categories of distinction, as he put it, to assist in determining the
H planning unit which should be considered in deciding whether there has been a material change of use. I quote those passages of most relevance:

> First, whenever it is possible to recognise a single main purpose of the occupier's use to his land to which secondary activities are incidental or ancillary, the whole unit of occupation should be considered . . .

*Editor's note: Now reported at [1991] 1 PLR 87.
†Editor's note: See [1991] 1 PLR 87 at p 95H.
§Editor's note: See [1991] 1 PLR 87 at pp 90H to 91A.

> But, secondly, it may equally be apt to consider the entire unit of A
> occupation even though the occupier carries on a variety of activities and it is
> not possible to say that one is incidental or ancillary to another . . .
>
> Thirdly, however, it may frequently occur that within a single unit of
> occupation two or more physically separate and distinct areas are occupied
> for substantially different and unrelated purposes. In such a case each area
> used for a different main purpose (together with its incidental and ancillary
> activities) ought to be considered as a separate planning unit . . . B
>
> It may be a useful working rule to assume that the unit of occupation is the
> appropriate planning unit, unless and until some smaller unit can be
> recognised as the site of activities which amount in substance to a separate use
> both physically and functionally.

Mr Village challenges the inspector's decision at para 9 of his decision letter
on three grounds, that is to say: first, he did not address his mind to whether
the garage building was a separate planning unit; second, he should have C
determined that it was a separate planning unit by reason of its location at the
end of the rear garden area and of the fact that it was self-contained and
therefore there was a material change of use and accordingly development
within section 55(1); and, third, he failed to give any reason for finding that the
dwellinghouse and garage were part of the same planning unit.

If, as Mr Village submits, there were two separate planning units and a
material change of use, he argued that the inspector should have gone on to D
consider within the terms of section 55(2)(d) whether, nevertheless,
development was not involved because the use of the garage was incidental to
the enjoyment of the dwellinghouse as such. On that second stage of the
correct approach he submits that the inspector should have found that, the
garage building with its residential annex being a separate planning unit, its
use as living accommodation was not incidental to the enjoyment of the E
principal dwellinghouse as such. On this question Mr Village referred to the
judgment of Sir Graham Eyre QC (sitting as a deputy High Court judge) in
Emin v *Secretary of State for the Environment* [1989] JPL 909 at p 913, where
the learned deputy judge said that "the word 'incidental' connoted an element
of subordination in land use terms in relation to the enjoyment of the dwelling-
house itself". Reliance was also placed upon a passage in the judgment of
Farquharson LJ in the case of *Wallington* (at p 6 of the transcript*), where he F
said: "In my judgment the word 'incidental' in this context means as Mr Laws
for the respondents puts it 'subordinate in land use terms to the enjoyment of
a dwellinghouse as a dwellinghouse'."

Lastly, Mr Village adopts as the correct approach which the inspector
should have followed but did not follow the views expressed by the Secretary
of State in a decision reported at [1987] JPL 144†. That decision was given on
an appeal against a determination which raised the question whether a G
proposal to erect a granny-annex building in the garden of a dwellinghouse
would be permitted development within class 1.3 of Schedule 1 to the Town
and Country Planning General Development Order 1977 as a building
required for a purpose incidental to the enjoyment of the dwellinghouse as
such. I read what appears to me to be the most relevant passages of that
decision: H

> Additions to the normal, basic, domestic living accommodation of a
> dwelling-house, such as bedrooms, which are normally to be expected as part
> and parcel of any dwelling's normal facilities, are not regarded, as a matter of
> fact and degree, as being "incidental" to the enjoyment of the dwelling-
> house as such for the purposes of Class 1.3; they are an integral part of the

*Editor's note: See [1991] 1 PLR 87 at p 97B.

†Editor's note: This refers to planning decision APP/K2230/G/84/38, September 23 1985.

> A ordinary residential use as a dwelling-house . . . The view is taken that the word "incidental", on the other hand, means something occurring together with something else and being subordinate to it. Accordingly, a purpose which is incidental to the enjoyment of a dwelling-house is distinct from activities which constitute actually living in a dwelling-house. Incidental purposes are regarded as being those connected with the running of the dwelling-house or with the domestic or leisure activities of the persons living
> B in it, rather than with the use as ordinary living accommodation. Similarly, . . . the Department's present view is that the use of an existing building in the garden of a dwelling-house for the provision of additional bedroom accommodation is not now to be regarded as being "incidental" to the enjoyment of a dwelling-house as such for the purposes of section 22(2)(*d*)

— that is a reference to the Town and Country Planning Act 1971 —

> it merely constitutes an integral part of the main use of the planning unit as a
> C single dwelling-house and, provided that the planning unit remains in single family occupation, does not therefore involve any material change of use of the land; in those circumstances it is now considered that there is therefore no need to rely on section 22(2)(*d*).

I have summarised Mr Village's submissions at some length because of the concern that the council is said to have at the implications of the inspector's decisions for other similar proposals, but I am bound to say that I do not
D accept that the decision has implications for proposals which are not identical on the facts. In considering those submissions I have in mind throughout the precise nature of the change of use for which planning permission was sought, that is to say, from "detached garage to private living accommodation to be used in conjunction with existing dwellings (in short 'a granny annex')".

This is ordinary language and I have no difficulty in understanding the
E proposal. The provision of such accommodation is common place.

It is not necessary that the inspector should have expressed himself as adopting a two-stage approach to the issue he had to decide if he in fact decided such questions as were required to be answered. The first was whether the application involved a material change of use. That turned on whether a separate planning unit would be created. In my judgment, the inspector addressed and decided that question. Para 9 of his decision letter must be read
F with the department's earlier letter, which conveyed his initial views and to which the inspector referred in his decision letter. The language which he there used may not be as precise as might reasonably be expected from a judge of this court, but I have no doubt that the inspector is there expressing his conclusion, albeit then provisionally, that the garage with the living accommodation would remain part of the same planning unit as the dwellinghouse. That was a finding of fact and degree and the council did not dissent from it in their letter,
G as the inspector observed in his decision letter. His reasons for that decision are there stated adequately and intelligently.

I can find no error of law in that decision. Mr Village submits that the inspector should have found that the garage with its living accommodation was a separate planning unit. The decision was, however, one of fact and degree. There was evidence on which the inspector could reasonably have
H decided as he did. Mr Village stressed the second respondent's representation that the accommodation gave the "occupant the facilities of a self-contained unit", although "intended to function as an annexe only with the occupant sharing her living activity in company with the family in the main dwelling". In so far as that was a material fact it was for the inspector to weigh it. I have no reason to conclude that he did not. As a material fact I do not think it has on the facts of this case the significance that Mr Village attributes to it. In the end it amounts to no more than the fact that the elderly relative to be accommodated

would have her own bedroom, bathroom and, I assume, lavatory, small kitchen, somewhere to sit and her own front door. To that extent she will be independent from the rest of her family. I find no reason in law why such accommodation should consequently become a separate planning unit from the main dwelling. So to conclude would suggest that the removal of any element of what was to be provided in this case, for example, the bathroom, such that the relative would to that extent be dependent on the facilities in the main house, would maintain that living accommodation in the same planning unit. There is, in my judgment, no reason in law why that should be so. A B

It follows that, having decided the first question arising on the two-stage approach required in the way he did, the second stage did not fall for determination. It is therefore unnecessary for me to make any judgment on the argument put to the court on the second of the two stages and I do not do so. Those are matters best left to be dealt with if and when they are in point. C

When the inspector came to consider the condition attached to the April 1989 permission he decided at para 13 of the decision letter:

> The condition as drafted is also self-contradictory since it purports to limit the use to purposes incidental to the enjoyment of the dwelling and then proceeds to exclude one such use — namely use as a residential annexe. In so far as it also seeks to preclude any separate industrial commercial or business use the condition is in my view unnecessary since any such separate uses would constitute a material change of use and require planning permission. D

Mr Village challenges the first part of this paragraph. He submits that the condition was not unnecessary or self-contradictory since use as a residential annex cannot be equated with use for purposes incidental to the enjoyment of the dwellinghouse.

Mr Sankey for the Secretary of State does not dispute that the reason given by the inspector for the condition being self-contradictory cannot be supported. If, as the inspector found, a garage with its living accommodation was part of the same planning unit as the dwellinghouse and provided that the planning unit remains in single family occupation, no material change of use is involved and no question arises of the use of the garage being used for a purpose incidental to the enjoyment of the dwellinghouse as such. Mr Sankey also submits that the inspector would have been right, however, to have found that the condition contradicted the planning permission to which it was attached and that his finding that it was unnecessary was a matter of planning judgment. E F

I accept Mr Sankey's submissions, which to an extent concedes Mr Village's point. Since, in my judgment, the inspector came to a correct conclusion although for the wrong reasons, I am not satisfied that the court is necessarily justified in quashing the decision on that limited ground. In any event I am satisfied that there is neither need nor purpose in the exercise of my discretion to quash it on that ground and to send it back for redetermination. Accordingly, I do not quash it. In the result the application is dismissed. G

Application dismissed.

Carpets of Worth Ltd *v* Wyre Forest District Council

COURT OF APPEAL
Purchas, Taylor and Beldam LJJ
March 12 1991

A **Alteration to green belt boundary — Land outside green belt in previous statutory development plan — Recommendation by inspector to include land in green belt accepted for local plan — Whether alteration of boundary of green belt requires "exceptional circumstances" — Meaning of Circular 14/84**

The appellants own 13 acres of open land at Puxton Lane, Kidderminster. B This land, together with further open land, a river and a canal, formed a "wedge" of undeveloped land between two areas of residential development. By amendment no 22 to the 1975 development plan their land was excluded from the green belt. The draft local plan published in May 1986 did not include the land in the green belt but, following a request by the county council, the respondent local planning authority altered the boundary to include it. The appellants' objection to the inclusion of their land in the green belt was not C accepted by the inspector who reported on the plan. His recommendation that the local plan should be amended to include the land in the green belt was accepted by the respondent local planning authority. The appellants challenged the validity of the Wyre Forest urban areas local plan in so far as it included their land in the green belt under section 244 of the Town and Country Planning Act 1971 (section 287 of the Town and Country Planning Act 1990). Otton J dismissed the application, (1990) 61 P&CR 57.

The appellants appealed from the decision of Otton J submitting that D neither the inspector nor the respondents in approving his recommendations had had regard to ministerial policy in Circular 14/84, para 3(*a*): ". . . detailed Green Belt boundaries defined in adopted local plans or earlier approved development plans should be altered only exceptionally." The respondents conceded that there were no exceptional circumstances.

Held The appeal was allowed.

E The respondents were not bound by ministerial policy in a circular or a PPG but were required to have regard to these documents: see p 88G. It was common ground that neither the inspector nor the respondents had had regard to Circular 14/84. The extension of the green belt to include the appellants' land was therefore an alteration to an earlier development plan. The general concept of the advice in the circulars is that once a green belt has been established and approved as a result of the normal statutory process it must F require exceptional circumstances rather than general planning concepts to justify an alteration: see pp 93-94. This accords with the plain reading of the words "altered" and "alteration" in para 3(*a*) of the circular: see p 94E.

Case referred to in the judgments
Gransden (EC) & Co Ltd v *Secretary of State for the Environment* (1985) 54 P&CR 86; [1986] JPL 519

G **Appeal against decision of Otton J**
This was an appeal against the decision of Otton J (March 7 1990) who dismissed an application by the appellants, Carpets of Worth Ltd, under section 244 of the Town and Country Planning Act 1971 (section 287 of the Town and Country Planning Act 1990) to challenge part of the local plan made by the respondents, Wyre Forest District Council.

Brian Ash QC and Joseph Harper (instructed by Marriott Harrison Bloom & Norris) appeared for the appellants, Carpets of Worth Ltd.

Christopher Wilson-Smith QC and Philip Mott (instructed by the solicitor to A
Wyre Forest District Council) appeared for the respondents.

The following judgments were delivered.

PURCHAS LJ: This is an appeal by Carpets of Worth Ltd ("Worth") from an order of Otton J made on March 7 1990 dismissing their application under section 244 of the Town and Country Planning Act 1971 (section 287 of the Town and Country Planning Act 1990). Worth seek to challenge part of the B
local plan made by Wyre Forest District Council which affected their land at Puxton Lane, Kidderminster. The appeal raises a single issue of importance touching upon the effect to be given to a provision of Circular 14/84, entitled *Green Belts*, issued by the Secretary of State for the Environment, to whom and to whose predecessors in office I shall refer as "the minister". It is common ground between Mr Christopher Wilson-Smith QC, who appeared C
for the council, and Mr Brian Ash QC, who appeared for Worth, that if the construction to be given to para 3(*a*) of that circular is that for which Mr Ash contends, this appeal succeeds but if the construction for which Mr Wilson-Smith contends succeeds, then the appeal must fail.

The position as thus agreed by counsel would appear happily to restrict the area for consideration by the court. It is accepted by both parties that in the exercise of their function as a local planning authority the council must have D
regard to "the provisions of circulars and similar documents called planning policy guidance (PPGs) issued by the minister". These documents announced the policy of the minister on the particular topics to which the circulars or PPGs were directed. The statutory basis for the proposition that these documents must be considered is to be found in section 11(9) of the 1971 Act:

> In formulating their proposals in a local plan the local planning authority E
> shall secure that the proposals conform generally to the structure plan as it stands for the time being (whether or not it has been approved by the Secretary of State) and shall have regard to any information and any other considerations which appear to them to be relevant, or which may be prescribed, or which the Secretary of State may in any particular case direct them to take into account.

"Prescribed" means prescribed by regulations made by the Secretary of F
State under the Act (see section 290(1)). The expression "Directions in any particular case" has not been defined either statutorily or by authority and there must be some doubt as to whether it can embrace circulars or PPGs. For the purpose of this judgment I shall assume that it does not. The current views of the minister as expressed in the circulars and the PPGs can only be material to the duties upon the council under section 11(9) as being "any information and any other considerations which appear to them to be relevant". G

Prior to the coming into effect of the code provided by the 1971 Act planning authorities in the counties (ie the county councils) had the responsibility of preparing development plans for their respective areas. These descended to detail only as to boundaries and areas but were supplemented by local town plans where appropriate. The development plans, *inter alia*, defined where the boundaries should be drawn between the green H
belt areas and areas zoned for other types of development. The green belt policy and the related development control policies were introduced in 1955.

Under provisions of Part II of the 1971 Act it was the duty of the local planning authority, which in the present case was Hereford and Worcester County Council, to prepare a structure plan. The statutory provisions requiring the appropriate survey, consultation etc leading to the submission of the draft structure plan by the county council to the minister for approval or

A rejection are to be found in sections 6 to 9 of the 1971 Act. Section 10 dealt with the subsequent alteration of structure plans once approved. The provisions relating to the preparation of local plans are contained in sections 11 to 15 of the 1971 Act. The preparation of a local plan was generally speaking a voluntary matter depending upon the decision of a local planning authority, which could be either the county council or a district council within the county: see section 3 of the 1971 Act.

B Worth's land at Puxton Lane, Kidderminster, was effectively divided into two sections. The southern part of about seven acres was at all material times occupied by their factory and warehouse premises. The northern part, consisting of some 13 acres or thereabouts, was open land lying immediately to the west of the River Stour. This appeal is concerned solely with the northern part. On the east side of the River Stour, speaking in very rough terms, lay
C further open ground between the river and the Staffordshire and Worcestershire canal, and further to the east was urban open development such as playing fields etc, before the residential development of Kidderminster, even further to the east, was reached. There was also to the west of the site residential development of Kidderminster. Again speaking in very general terms, the open-type development, the river, canal and other land formed an "inlet" of undeveloped land between the eastern and western
D residential developments.

This has been described as a "wedge" and I shall refer to it as such hereafter in this judgment. The word "wedge" although used in planning parlance is not a term of art and has no statutory significance.

The development plan had been prepared by Worcestershire County Council under the pre-1971 legislation. It showed a green belt area running generally north of Kidderminster. It was subject to a number of amendments.
E The relevant amendment for this appeal was amendment no 22 which was approved in 1977. This showed the southern boundary for the green belt to the north of Kidderminster as running east to west in a general line tangential to the northern limits of the existing residential developments in the western and eastern parts, but it did not include the wedge. It ran across the northern base of the wedge. Worth's land was, therefore, excluded from the green belt.

F In November 1982, the county council published draft proposals for a green belt local plan covering the whole county. This included a map ("map 35") showing an alteration locally to the pre-existing green belt boundary to the north of Kidderminster, which had been approved in amendment no 22. Generally speaking it extended the green belt to include the wedge. It was described as "proposal 24", which read:

> Location: The Stour Valley between the Franche area and Broadwaters area.
> G It is proposed that the open land on either side of the River Stour, with the exception of the areas already laid out as urban open spaces, should be included in the Green Belt. To protect this potentially attractive open area, it is considered that the areas already laid out for urban open space purposes are adequately protected from building development.

Nothing seems to have come of this proposal. Perhaps it was overtaken in the
H general process of preparing and submitting the draft structure plan for approval by the Secretary of State. This was finally approved in September 1985.

In accordance with their duties under the 1971 Act the county council prepared a structure plan. This was published in October 1985 as approved by the Secretary of State. A key diagram which accompanied the plan when submitted for approval showed only the general position of the green belt and is referred to in para 9.6 of the Secretary of State's letter:

> On the key diagram which accompanied the submitted structure plan, the County Council had shown the general area of the Green Belt between Droitwich and Worcester, as extending as far west as the River Severn. In the approved Worcestershire Structure Plan the extent of the Green Belt lay generally between the M5 Motorway to the east and the Birmingham to Worcester railway line to the west. The Secretary of State proposes to confine the Green Belt generally to the area in the approved Structure Plan. The Secretary of State is committed to the preservation of the Green Belt and considers that for Green Belts to maintain their credibility, once their general extent has been approved as part of the Structure Plan for the area, they should be altered only in exceptional circumstances. He is aware that precise Green Belt boundaries remain to be determined in this area and that this exercise, which should be done through the Local Plan process, may involve some adjustment at the margins of the Green Belt as shown in the Structure Plan. However, he considers that the proposed alteration of the extent of the Green Belt in the plan as submitted goes further than a marginal adjustment. He does not consider that the basic principle of Green Belt (preventing the coalescence of settlements) is at risk for the whole of the area.

A B C

On May 28 1986 the council published a draft of their proposed local plan. This did not follow the proposals in the draft county council plan so as to include the applicants' land in the proposed green belt. This prompted representations from the county council. In the light of these and other objections the council published a further draft plan, which became part of the statutory consultative procedure. On this plan the green belt was projected southwards to include undeveloped land in the wedge itself. This proposal, known as "proposal 3", formed part of a schedule of proposed changes to the green belt boundary published together with the draft local plan. The comment against the change, which was described as change from "Part Public Open Space, Part White Land, Part Residential to Green Belt", was explained in the following manner:

D E

> To safeguard the open valley and provide an area for informal recreation.

This contrasted with the description in the draft local green belt plan proposed by the county council in proposal 24 which I have already recited. The reasons for the proposal were:

F

> (2) Reasons.
> To protect this potentially attractive open area. It is considered that the areas already laid out for urban open space purposes are adequately protected from building development.

Para 2.5 of the approved county structure plan defined the policy for the green belt in the County of Hereford and Worcester:

G

> (a) To prevent further growth of the conurbation into the countryside;
> (b) to limit the expansion of built-up areas in the Green Belt area in order that neighbouring towns and villages will not merge with one another; and
> (c) to safeguard the area of open countryside in order to take account of the interests of agriculture and to provide a source of informal recreation and enjoyment for the inhabitants of the area and neighbouring built-up areas. The rural character of the Green Belt will therefore be retained, protected and when the opportunity arises, enhanced.

H

Worth objected to proposal 3 on the basis that the proposed extension of the green belt to include the wedge did not serve to prevent further growth of the conurbation into the countryside or to prevent neighbouring towns and villages merging into one another. Therefore purposes (a) and (b) were not valid reasons for this alteration. Furthermore, Worth did not accept that purpose (c) could stand on its own as a reason for a new incursion into other

A planning purposes by the green belt. A local public inquiry was accordingly held under the provisions of section 13 of the 1971 Act.

At the public inquiry Worth through their advisers proposed that so far as the land to the north of their factory area was concerned there should be three different types of development. These included residential development, some public open space and amenities. Having set out in his report the principal objections and the council's proposals the inspector at para 300,
B under the heading "*Green Belt Boundary Proposals*", said:

> 300. I have already indicated that objections relating to the proposed Green Belt boundary but which concern specific sites have been covered elsewhere in this report. In this section I consider the Council's Schedule of proposed principal changes to the boundary as set out in Plans 5 to 11 and in Appendix 3 to the Written Statement.
> C *Change Three*: Stour Valley. Puxton.
> This proposal will take in part of objection Sites Nos. 5 and 7 and part of Site No. 8. These areas have been recommended for inclusion in the Green Belt. This change is therefore accepted.

The report and recommendations of the local plans working party on the report into objections to the local plan came before the planning and highways committee of the council on September 28 1988, when para 300 of the report,
D which has just been cited, was accepted without change from the inspector's recommendations.

Worth now challenges the adoption of this part of the inspector's report and its inclusion in the local plan on the grounds that neither the inspector, nor the council in approving his recommendations without more consideration, had regard to ministerial policy relating to the treatment of green belt proposals in local plans, as disclosed particularly in Circular 14/84 para 3(*a*). It is true that
E the inspector's report was considered at a meeting of the planning and highways committee on September 28 1988 — but there is no record of any consideration of exceptional circumstances justifying the alteration of the boundary of the green belt. Indeed Mr Wilson-Smith does not suggest that such exceptional circumstances existed in any event.

It is necessary at this stage to refer to some of the circulars and PPGs. Before
F doing so, however, notwithstanding the accord reached between Mr Wilson-Smith and Mr Ash, I must consider the status of these documents. They are not issued under statutory authority. "Prescribed" considerations involve regulations made by the Secretary of State under section 287 of the 1971 Act and are therefore subject to resolution of each House of Parliament. Ministerial circulars as published or as summarised in PPGs have therefore no formal statutory force and should therefore not be treated as such for any
G purpose. This includes, in my judgment, the manner in which they should be construed and/or applied. They constitute announcements of the current ministerial planning policy. The only statutory obligation upon the local planning authority is "to have regard to them". They are in no way bound by them. This appeal can be based only on the ground that the council did not have regard to a relevant circular or PPG.

H The concept of green belts was first introduced in 1955 consequent upon a statement by the minister in the House of Commons on April 26 of that year.

Circular 42/55 dated August 3 1955, after referring to the statement in the House, stated:

> 1 . . . I am directed by the Minister of Housing and Local Government to draw your attention to the importance of checking the unrestricted sprawl of the built-up areas, and of safeguarding the surrounding countryside against further encroachment.

2 He is satisfied that the only really effective way to achieve this object is by the formal designation of clearly defined Green Belts around the areas concerned.

3 The Minister accordingly recommends Planning Authorities to consider establishing a Green Belt wherever this is desirable in order:

(*a*) to check the further growth of a large built-up area;

(*b*) to prevent neighbouring towns from merging into one another; or

(*c*) to preserve the special character of a town.

4 Wherever practicable, a Green Belt should be several miles wide, so as to ensure an appreciable rural zone all round the built-up area concerned.

5 Inside a Green Belt, approval should not be given, except in very special circumstances, for the construction of new buildings or for the use of existing buildings for purposes other than agriculture, sport, cemeteries, institutions standing in extensive grounds, or other uses appropriate to a rural area.

. . .

8 In due course, a detailed survey will be needed to define precisely the inner and outer boundaries of the Green Belt, as well as the boundaries of towns and villages within it. Thereafter, these particulars will have to be incorporated as amendments in the Development Plan.

No 9 is not relevant.

Circular 50/57 — *Green Belts* dated September 19 1957:

5. The definition of a long-term boundary for development may involve detailed adjustments (either inwards or outwards) in the boundary of the area already allocated on a Town Map. Where land allocations are to be deleted or additional land allocated for development within the Plan period, the adjustments can be included in the same submission as the Green Belt proposals.

6. There may be some pockets of land, between the town and the Green Belt, which are not to be developed within the present Plan period but which could be developed later without prejudice to the Green Belt. It would be misleading to allocate such areas now, but to include them in the Green Belt for the time being might give rise to difficulties and undermine public confidence in the Green Belt at a later date if it were then decided to allocate the land for development. Such areas may well be left as pockets of "white" land. They are then bound to be especially attractive to developers and it will be desirable to set out in the Written Statement the authority's policy for such areas in order to make it clear that they are not available for development at the present time.

Circular 14/84 — *Green Belts* dated July 4 1984:

1. The Government continues to attach great importance to Green Belts, which have a broad and positive role in checking the unrestricted sprawl of built-up areas, safeguarding the surrounding countryside from further encroachment and assisting in urban regeneration. There must continue to be a general presumption against inappropriate development within Green Belts. The Government reaffirms the objectives of Green Belt policy and the related development control policies set out in Ministry of Housing and Local Government Circular 42/55.

2. Structure plans have now been approved for most parts of the country and these identify the broad areas of the Green Belt. Detailed Green Belt boundaries are now being defined in local plans and in many cases these are based on Green Belt areas defined in earlier development plans approved prior to the introduction of structure and local plans. This process of local plan preparation is continuing and this circular includes advice on the definition of detailed Green Belt boundaries in local plans.

3. The essential characteristic of Green Belts is their permanence and their

A protection must be maintained as far as can be seen ahead. It follows from this that:

> (*a*) Once the general extent of a Green Belt has been approved as part of the structure plan for an area it should be altered only in exceptional circumstances. If such an alteration is proposed the Secretary of State will wish to be satisfied that the authority has considered opportunities for development within the urban areas contained by
> B and beyond the Green Belt. *Similarly, detailed Green Belt boundaries defined in adopted local plans or earlier approved development plans should be altered only exceptionally.*

(Emphasis provided.)

Para (*b*) is not relevant.

It remains only to refer to the relevant PPG. There were a series of these
C published in draft form towards the end of 1987 and formally published in January and February 1988. There were in fact 10 in number; but at this stage I need only refer to one of them.

Planning Policy Guidance Note 2 was headed *Green Belts* and provided so far as relevant as follows:

> **Purposes of Green Belts**
>
> 4. Green Belts have five purposes:
> D — to check the unrestricted sprawl of large built-up areas;
> — to safeguard the surrounding countryside from further encroachment;
> — to prevent neighbouring towns from merging into one another;
> — to preserve the special character of historic towns; and
> — to assist in urban regeneration.
>
> 5. Green Belts also have a positive role in providing access to open
> E countryside for the urban population. Such access may be for active outdoor sports or for passive recreation. Outdoor leisure pursuits are likely to occupy an increasing proportion of the Green Belt if, as currently expected, the land needed for food production decreases. . . .
>
> **Designation of Green Belts**
>
> 7. The essential characteristic of Green Belts is their permanence and their protection must be maintained as far as can be seen ahead.
>
> F 8. Green Belts are established through development plans. Their general extent has now been fixed through the approval of structure plans and many detailed boundaries have been set in local plans and in old development plans.
>
> 9. Once the general extent of a Green Belt has been approved it should be altered only in exceptional circumstances. If such an alteration is proposed the Secretary of State will wish to be satisfied that the authority has considered opportunities for development within urban areas contained by
> G and beyond the Green Belt. Similarly, detailed Green Belt boundaries defined in adopted local plans or earlier approved development plans should be altered only exceptionally. Detailed boundaries should not be amended or development allowed merely because the land has become derelict. On the outer edge of a Green Belt, readily recognisable features such as roads, streams or belts of trees should be used to define the boundaries.
>
> 10. Where detailed Green Belt boundaries have not yet been defined, local
> H planning authorities are urged to complete this task. *It is necessary to establish boundaries that will endure and they should be carefully drawn so as not to include land which it is unnecessary to keep permanently open. Otherwise there is a risk that encroachment on the Green Belt will have to be allowed in order to accommodate future development.*
>
> 11. When local planning authorities prepare new or revised structure and local plans, *any proposals affecting Green Belts should be related to a time scale which is longer than that normally adopted for other aspects of the plan. They should satisfy themselves that Green Belt boundaries will not need to be*

> *altered at the end of the plan period. In some cases this will mean safeguarding land between the urban area and the Green Belt which may be required to meet longer term development needs.*

(Emphasis provided.)

This appeal therefore depends upon whether or not para 3(*a*) of Circular 14/84 applies to the decision of the council to adopt the inspector's recommendation that the wedge should be included within the boundaries of the green belt shown on their local plan. It is common ground that neither the inspector nor the planning and highways committee of the council "had regard to this circular." It is also common ground that they had a statutory obligation to do this under section 11(9) of the 1971 Act. Although the local authority is not bound by the policy circulars, it should observe them and depart from them only if there are clear reasons, which should be stated, for so doing: see *Gransden (E C) & Co Ltd* v *Secretary of State for the Environment* (1985) 54 P&CR 86.

There was an "earlier approved development plan" which excluded the wedge from the green belt (amendment no 22 to the development plan prepared by Worcestershire County Council already mentioned). The extension of the southern boundary of the green belt to include the wedge was therefore an alteration to the boundaries defined in an earlier approved development plan. The central issue was whether "exceptionally" related to any alteration or only to alterations to the boundary the effect of which was to diminish the area of the green belt in the immediate area of the variation.

Otton J considered this question at (1990) 61 P&CR 57 at pp 62-63:

> Thus when one turns to para 3 (of Circular 14/84), one sees that the Secretary of State reiterates the essential characteristics of green belts and then states:
>
> > "Once the general extent of a Green Belt has been approved as part of the structure plan for an area it should be altered only in exceptional circumstances."
>
> This passage is not strictly relevant to this case, but it none the less indicates that the general extent is determined and settled in the structure plan and once it has been so determined it shall then (ie in the future) be altered only in exceptional circumstances. In short, the Secretary of State contemplates a prospective and not a retrospective approach. Thus, when one reaches the next relevant sentence, "Similarly, detailed Green Belt boundaries defined in adopted local plans . . . should be altered only exceptionally", the meaning is clear. Once the detailed boundaries have been defined and adopted in a local plan, then (ie after adoption) they should be altered only in exceptional circumstances.
>
> I have of course omitted the words "or earlier approved development plans". The reinsertion of those words into that sentence does not, in my judgment, alter the construction I have placed upon the sentence as a whole. These words, as I read them, refer to the rare situations where there is no local plan for the area or, if there is a local plan, it expressly adopted the boundaries defined in an earlier approved plan.

With respect to Otton J I find it difficult to see how the exercise of omitting the words "or earlier approved development plans" assists in the construction of the passage as a whole. "Adopted local plans" are local plans which have been prepared under sections 11 to 14 of the 1971 Act. These can be altered subsequently under the provisions of section 15 of that Act. "Earlier approved development plans" clearly relates to development plans prepared and approved under section 55 of the 1947 Act, including amendments under section 56 of that Act. The two types of plans to which para 3(*a*) refers are quite different classes of document and are prepared under different statutes. Of course, the plan with which this appeal is concerned falls within the latter

A rather than the former category. In construing para 3(*a*) one must look at it as a whole.

The first sentence of para 2 of Circular 14/84 recorded that structure plans had by then been approved for most parts of the country and that these identified the broad areas of green belt. The remaining part of para 2 addressed itself to the question of local plans in which detailed boundaries of
B green belts had already been shown, or were about to be shown, stating that in many cases these were based on green belt areas defined in earlier relevant plans approved prior to the introduction of structure and local plans. Here again the relevant plans referred to would include development plans. Returning to para 3(*a*), the first point that the paragraph makes is that once the "general extent" of a green belt has been approved this should be "altered" only in exceptional circumstances. It is to be noticed that the
C paragraph does not say that it should be "reduced" only in exceptional circumstances. The word "altered" is quite unqualified. What, then, is the significance of the word "similarly" with which the last sentence opens? It must refer to two categories of cases, namely where local plans had already been adopted or where earlier approved development plans were in place. I see no mandate for treating the two alternative positions differently. The boundaries of existing green belts in structure plans should not be altered
D either way except in exceptional circumstances nor should adopted local plans be treated any differently. Mr Wilson-Smith argued that the word "alteration" as used in para 3 of the circular means "alteration which results in diminishing the area of the green belt". He justified the implication of these words by pointing to the policy of Circular 14/84, which refers only to the necessity of "preserving" the green belt from encroachment, usually by buildings, residential and industrial, both within the green belt or by erosion
E along the boundaries. He referred by way of example to "safeguarding the surrounding countryside from further encroachment . . ." (para 1):

> There must continue to be a general presumption against inappropriate development within Green Belts

and (para 3):

F
> The essential characteristic of Green Belts is their permanence and their protection must be maintained . . .

These statements merely repeat and reinforce Circular 42/55, to which I have already referred. The context in which the statement in the House was made and the circular issued was a purely negative one, ie to prevent urban sprawl. It is not surprising, therefore, that the emphasis is on the restriction of
G alterations, the effect of which would be to erode or diminish the extent of green belts. Mr Wilson-Smith submitted that there was no need to give protection against extensions of the green belt. I must return to consider this proposition in a little more detail later. It was not the basis upon which Otton J formed his judgment, (1990) 61 P&CR 57 at p 63:

> In short I reject the construction advanced by Mr Harper [counsel for the applicant]. Para 3(*a*), read in the context of Circular 14/84 as a whole, does
H not mean that once a boundary had been defined and settled under an earlier development plan it could only be changed in a later local plan in exceptional circumstances.
>
> In reaching this decision, I take into account part of para 9.6 from the Hereford and Worcester County Structure Plan written statement, which states:
>
> "The Secretary of State is committed to the preservation of the green belt and considers that for green belts to maintain their credibility, once their

> general extent has been approved as part of the structure plan for the area, **A** they should be altered only in exceptional circumstances."
>
> I interpolate that the words "once . . ." through to "exceptional circumstances" reiterate word for word the first sentence of para 3(*a*) of Circular 14/84. The next sentence appears to confirm my construction:
>
> > "He is aware that precise green boundaries remain to be determined in this area and that this exercise, which should be done through the local plan process, may involve some adjustment at the margins of the green belt as **B** shown in the structure plan".
>
> In my judgment the alteration of the applicants' boundary as a result of the representation by the county council amounted to no more than an adjustment at the margin of the green belt.

I hope that it does not do an injustice if I say that Mr Wilson-Smith did not embrace the suggestion that the inclusion of the wedge within the green belt could be described as an adjustment at the margin with any degree of **C** enthusiasm. The main plank of his argument was that alterations to the boundary of the green belt, which had the effect of extending the green belt, did not require justification by exceptional circumstances. To support this argument the respondents had served a notice under RSC Ord 59 r 6(1)(b) to the following effect:

> 1. That exceptional circumstances were not required under Circular 14/84 **D** or Planning Policy Guidance (PPG2) issued by the Department of the Environment where the alteration to the green belt was by way of extension rather than relaxation.
>
> 2. That the process of producing a new Local Plan was itself an exceptional circumstance.

Para 2 of the notice was not relied upon by Mr Wilson-Smith, rightly in my judgment. It could not be substantiated on any basis. **E**

Mr Ash submitted that para 3(*a*) of Circular 14/84 was in plain terms and that proper planning required consideration of the interests of all parties involved, not the least those interested in development. The effect of the green belt was to sterilise the area for development, other than the limited number of specified purposes, and would inhibit proper and orderly development. Therefore, once the boundaries of the green belt in a particular area had been defined, developers and others interested were entitled to protection against **F** change to the limited degree that the boundaries would be altered only in exceptional circumstances. Besides what must be assumed to be the deliberate use of the word "alter" in para 3(*a*) as I have already observed in my judgment, Mr Ash has support for his submissions to be drawn from a consideration of planning concepts as a whole.

The circulars and PPGs having no statutory authority can be viewed only as indicatory of current planning policy. Although the PPGs were issued over a **G** matter of months, the circulars were published over a period of years. One can detect over the passage of time shifts in emphasis disclosed in the latter as between "development" on the one hand and "conservation" on the other. In the circumstances, I think that Mr Wilson-Smith was justified in submitting that these documents should not be treated as statutes in the process of construing what they mean. In my judgment, it is legitimate to look to the **H** purpose of planning policy as a whole and take into account the particular contemporaneous context in which a particular document was published. I consider that the key may well lie in the underlying concept of PPG 1, *General Policy and Principles*, namely that, where possible, planning policies should encourage development:

> **The presumption in favour of development**
>
> 15. The planning system fails in its function whenever it prevents, inhibits

> or delays development which can reasonably be permitted. There is always a presumption in favour of allowing applications for development, having regard to all material considerations, unless that development would cause demonstrable harm to interests of acknowledged importance. Except in the case of inappropriate development in the Green Belt the developer is not required to prove the case for the development he proposes to carry out; if the planning authority consider it necessary to refuse permission, the onus is on them to demonstrate clearly why the development cannot be permitted.

The zoning of any particular area as a green belt sterilises that area except for the limited category of purposes listed in para 5 of Circular 42/55. It therefore provides a kind of planning blight and must be an exception to PPG 1 para 15, the area of which should not be extended unless it is necessary for the purposes of the green belt as defined in PPG 2.

As it directly prejudices landowners in the otherwise proper development of their land, an extension to the green belt should not be brought into effect unless it can be justified directly by those purposes for which the green belt is designed. There must, therefore, be an inhibition in extending a green belt so as to avoid sterilising unnecessarily neighbouring land (see the provisions of paras 5 and 6 of Circular 50/57 set out earlier in this judgment) just as much as reductions in the boundaries of the green belt, which would prejudice the purposes of that green belt, must also be made only in exceptional circumstances. On this basis I think that the general concept of the advice in the circulars is that once a green belt has been established and approved as a result of all the normal statutory processes it must require exceptional circumstances rather than general planning concepts to justify an alteration. Whichever way the boundary is altered there must be serious prejudice one way or the other to the parties involved.

This accords with a plain reading of the words "altered" and "alteration" in para 3(*a*) of Circular 14/84. Furthermore, the general tenor of Circular 14/84 is that once a green belt has been approved — either in an adopted local plan or, if a local plan has not been adopted under the 1971 legislation, in the next previously approved and adopted plan — alteration to boundaries of a green belt should be made only in exceptional circumstances. There are two obvious qualifications to this principle: first, if as a result of the supervening structure plan green belt boundaries shown in an earlier development become meaningless or anomalous; second, if the structure plan for the area concerned has not been approved, then none of the provisions of para 3(*a*) apply. The matter is considered in para 3(*b*), to which it is, at this stage, interesting to refer:

> Where detailed Green Belt boundaries have not yet been defined in earlier approved development plans or in adopted local plans — for example, where approved structure plans have extended the area of the Green Belt to include areas previously referred to as "interim" Green Belt — it is necessary to establish boundaries that will endure. It is especially important that these boundaries of Green Belts should be carefully drawn so as not to include land which it is unnecessary to keep permanently open for the purpose of the Green Belt. Otherwise there is a risk that encroachment on the Green Belt may have to be allowed in order to accommodate future development. If Green Belt boundaries are drawn excessively tightly around existing built-up areas it may not be possible to maintain the degree of permanence that Green Belts should have. This would devalue the concept of the Green Belt and also reduce the value of local plans in making proper provision for necessary development in the future.

Although, for the reasons already stated, the emphasis is on erosion of the green belt, this subparagraph indicates that the importance of unnecessary extension has not been overlooked.

Although there might have been a case to be made on the grounds that the A reasons given in proposal 3 to the draft local plan for extending the green belt were not justifiable in any event, this was not pursued by Mr Ash and it is not necessary for me to consider this position. I am satisfied that neither the inspector nor the planning authority "had regard" to para 3(*a*) of Circular 14/84. For the purpose of the appeal it was conceded by Mr Wilson-Smith that there were no "exceptional circumstances" which would justify the inclusion of the wedge into the green belt. I would allow the appeal and quash this part of B the local plan.

TAYLOR LJ: I agree that this appeal should be allowed for the reasons given by Purchas LJ. There was no dispute that the local planning authority had a duty to have regard to circulars and PPGs issued by the Secretary of State. The central issue was as to the meaning and effect of para 3(*a*) of Circular 14/84, C which provides as follows:

> 3. The essential characteristic of Green Belts is their permanence and their protection must be maintained as far as can be seen ahead. It follows from this that:
>
> (*a*) Once the general extent of a Green belt has been approved as part of the structure plan for an area it should be altered only in exceptional circumstances. If such an alteration is proposed the Secretary of State D will wish to be satisfied that the authority has considered opportunities for development within the urban areas contained by and beyond the Green Belt. Similarly, detailed Green Belt boundaries defined in adopted local plans or earlier approved development plans should be altered only exceptionally.

Without seeking to construe the circular as strictly as a statute, which would be inappropriate, I agree with Purchas LJ as to its purpose and effect. The E introductory words refer to the permanence as well as to the protection of the green belt as being essential. The word "altered" is used in para 3(*a*) rather than words such as reduced, encroached upon or breached. It is suggested that the second sentence is consistent only with alteration by reduction, the implication being that if there are opportunities for development in the urban areas the green belt should not be invaded for such development. However, it F could equally apply to exclude extension of the green belt if such extension would diminish unacceptably opportunities for development having regard to the opportunities available in the urban areas. This latter consideration is clearly contemplated in para 3(*b*), which states, *inter alia*:

> . . . it is especially important that these boundaries of Green Belts should be carefully drawn so as not to include land which it is unnecessary to keep permanently open for the purpose of the Green Belt. Otherwise there is a risk G that encroachment on the Green Belt may have to be allowed in order to accommodate future development. If Green Belt boundaries are drawn excessively tightly around existing built-up areas it may not be possible to maintain the degree of permanence that Green Belts should have . . .

In my view, the requirement of exceptional circumstances before altering the green belt was applicable to increasing as well as reducing it. The prejudice H to landowners and developers and the uncertainties which would be created if the green belt, once approved, could be extended other than exceptionally have been fully explained by Purchas LJ. It is true that the circular and PPG 2, which is to similar effect, do not bind the local planning authority, but they must have regard to them. They did not do so. Moreover, it is conceded that there were here no exceptional circumstances. Accordingly, in my view, the decision to include the wedge in the green belt was flawed.

A **BELDAM LJ:** In this case the appellants in an application to the High Court under section 244 of the Town and Country Planning Act 1971, challenge the validity of a proposal made by the respondent district council in formulating the Wyre Forest urban areas local plan on the ground that they did not have regard to considerations which the Secretary of State in the particular case directed them to take into account. The particular considerations are those B contained in circulars and planning policy guidelines relating to green belt areas. As Purchas LJ records in his judgment, the court has been asked to decide the case on the single issue of the construction to be given to a particular circular, Circular 14/84, issued by the Secretary of State and entitled *Green Belts*. It is said that if the construction contended for by the appellants is correct, their appeal should succeed but if the respondents' construction is right then the appeal must fail. I do not myself believe that this is the correct or C a realistic basis upon which to decide whether in any particular case a local planning authority have erred in formulating their proposals in a local plan. Nevertheless, under the constraint imposed by the manner in which the case was presented and in the light of the clear concession made on the respondents' behalf by Mr Wilson-Smith QC that the respondents' reasons for their decision would be inadequate if they were required to have regard to the Secretary of State's policy for settling the boundaries of a green belt area when D enlarging as well as when reducing the area contained within its boundaries, I agree that the appeal should be allowed.

Ministerial guidance clearly does not have the force of a statute or regulations and is not intended to be subjected to a process of legalistic interpretation. Equally, ministerial policy is something which a local planning authority are required under the Act to take into account in reaching the decision entrusted to them by Parliament. According to para 5 of PPG 2 E issued by the Department of Environment in January 1988, one justification for including land within a green belt is so that it may play a positive role in providing access to open countryside for the urban population. Such access may be for active outdoor sports or for passive recreation. The reason given in appendix 3 of the Wyre Forest urban areas local plan written statement for the inclusion in the green belt of the wedge of land, part of which belonged to the F appellants, was to safeguard the open valley and to provide an area for informal recreation, a purpose which, as it seems to me, was well within the justification to which I have referred.

The paragraphs in Circular 14/84 dated July 4 1984 and issued by the Department of the Environment upon which so much emphasis has been placed are in part reproduced, under the heading *Designation of Green Belts*, in paras 7 to 11 of PPG 2. If it were correct to view the ministerial guidance as G being subject to strict legal rules of construction, I would, for my part, have said that in para 3(*a*) the minister clearly had in mind alterations which consisted of allowing development in designated green belt areas. The fact that if such an alteration were proposed the Secretary of State would have to be satisfied that it could not be accommodated within urban areas contained by and beyond the green belt seems to me to support this. The word "similarly" later in para 3(*a*) is, in my view, used to indicate that the same criterion should H be applied to green belt boundaries defined in adopted local plans or earlier approved development plans as to those approved as part of a structure plan for an area. And although this paragraph in isolation might suggest that ministerial guidance is limited to development which reduces the area of the green belt, there is clearly other policy guidance for green belt boundaries in both the circular and in PPG 2. Circular 14/84 states at para 3(*b*):

> . . . It is especially important that these boundaries of Green Belts should be

> carefully drawn so as not to include land which it is unnecessary to keep A
> permanently open for the purpose of the Green Belt. Otherwise there is a risk that encroachment on the Green Belt may have to be allowed in order to accommodate future development. If Green Belt boundaries are drawn excessively tightly around existing built-up areas, it may not be possible to maintain the degree of permanence that Green Belts should have. This would devalue the concept of the Green Belt and also reduce the value of local plans in making proper provision for necessary development in the future. B

This guidance is reproduced in PPG 2 in para 10:

> . . . It is necessary to establish boundaries that will endure and they should be carefully drawn so as not to include land which it is unnecessary to keep permanently open. Otherwise there is a risk that encroachment on the Green Belt will have to be allowed in order to accommodate future development.

C

Such guidance cannot, I think, be confined to cases where detailed green belt boundaries have not yet been defined. In the context of ministerial policy as a whole it is, I think, intended to apply whether the boundaries of the green belt are being changed to reduce or to enlarge the area. This seems to me clear from the policy more fully set out in Circular 14/84. Thus, a local planning authority considering whether or not to alter the boundaries of a green belt by
the inclusion of additional land not previously contained within the boundary D
should at least take into account as part of ministerial policy whether that boundary is in the longer term defensible against pressure for development and whether to draw the proposed boundary may include land which it is unnecessary to keep permanently open for the purpose of the green belt. The reasons given by the respondents and the manner in which the appeal has been
resisted on their behalf seem to me to indicate that in adopting the E
recommendation of the inspector who conducted the local inquiry they did not have regard to this important aspect of ministerial policy.

I, therefore, agree that the appeal should be allowed.

Appeal allowed with costs here and below.

F

South Lakeland District Council *v* Secretary of State for the Environment and another

COURT OF APPEAL

Butler-Sloss and Mann LJJ and Sir Christopher Slade G

March 12 1991

Statutory duty under section 277(8) of 1971 Act — Proposed development neutral in affecting conservation area — Whether "preservation" has narrow meaning apt to include only development bringing positive contribution or wider meaning of bringing no harm — Wider meaning preferred

H

By a decision of July 13 1989 an inspector appointed by the appellant, the Secretary of State for the Environment, allowed an appeal on written representations of the second respondents, Carlisle Diocesan Parsonages Board, and granted outline planning permission for the erection of a new vicarage within the curtilage of the existing vicarage at Priest Lane, Cartmel, Cumbria. The application of the first respondents, South Lakeland District Council, to quash the Secretary of State's decision was allowed by Mr Lionel

Read QC (sitting as a deputy judge of the Queen's Bench Division — February 26 1990) on the ground that the inspector had not had proper regard to the statutory duty imposed on him by section 277(8) of the Town and Country Planning Act 1971 (section 72(1) of the Planning (Listed Buildings and Conservation Areas) Act 1990). The inspector had found that the proposed development did not enhance the conservation area; the deputy judge, in giving the word "preservation" a narrow meaning in that the proposed development did not bring a positive contribution to the area, decided that the statutory duty under section 277(8) had therefore not been discharged. The Secretary of State appealed.

Held The appeal was allowed.

The inspector found that neither the character nor the appearance of the conservation area would be harmed by the proposed vicarage, and the character of the area would remain preserved notwithstanding the development. In the absence of any authority binding on the court as to the meaning of "preserving" in section 277(8), a wide meaning should be given; the character or appearance of an area can properly be said to be preserved by development which would bring no harm: see p 101. The legislature did not intend the word "preserve" to have a narrow meaning as referring to development intended to have a positive contribution to preservation and to exclude development of a neutral character: see pp 102 and 104. The inspector had identified his statutory duty which he discharged. The present case was not within the sixth principle enunciated by Glidewell LJ in *Bath Society* v *Secretary of State for the Environment* [1991] 2 PLR 51: see p 106.

Cases referred to in the judgments

Bath Society v *Secretary of State for the Environment* [1991] 2 PLR 51, CA
Harrow London Borough Council v *Secretary of State for the Environment* (1989) 60 P&CR 525; [1990] 2 PLR 62
Steinberg v *Secretary of State for the Environment* (1989) 58 P&CR 453; [1989] 2 PLR 9; [1989] JPL 258

Appeal against decision of Mr Lionel Read QC

This was an appeal against a decision of Mr Lionel Read QC (sitting as a deputy judge of the Queen's Bench Division — February 26 1990) who had allowed an application of the first respondents, South Lakeland District Council, to quash the decision of the appellant, the Secretary of State for the Environment, who had allowed an appeal by his inspector against a decision of the first respondents to refuse an application for planning permission submitted by the second respondents, Carlisle Diocesan Parsonages Board.

John Laws and Ian Ashford Thom (instructed by the Treasury Solicitor) appeared for the appellant, the Secretary of State for the Environment.
Nigel Macleod QC and Jonathan Karas (instructed by Sherwood & Co) appeared for the first respondents, South Lakeside District Council.
The second respondents, Carlisle Diocesan Parsonages Board, did not appear and were not represented.

The following judgments were delivered. Cur adv vult

MANN LJ: This is an appeal from a decision of Mr Lionel Read QC sitting as a deputy judge of the Queen's Bench Division on February 26 1990. On that day he allowed an application by South Lakeland District Council under section 245 of the Town and Country Planning Act 1971 [section 288 of the Town and Country Planning Act 1990] and quashed a decision by an inspector appointed by the Secretary of State for the Environment dated July 13 1989. The basis of the learned deputy judge's decision was that the inspector had made an error in law in that he had not paid proper regard to the requirements of section 277(8) of the Act of 1971.

The Act of 1971 has now been replaced by the consolidating legislation of A
1990 and in particular by the Town and Country Planning Act 1990. The new provisions are not in substance different from the old, and for convenience I shall refer to the provisions in force at the time of the events giving rise to this appeal.

The events commence on July 20 1988 when the South Lakeland District Council received an application for planning permission by the Carlisle B
Diocesan Parsonages Board. It was an application for outline permission to erect a new vicarage within the curtilage of the existing vicarage at Priest Lane, Cartmel, Cumbria. The site is within the Cartmel Conservation Area which was designated in 1969. Cartmel is a village which is rich with archaeological, architectural, aesthetic and historic character. On September 12 1988 the application was rejected for reasons that:

> The proposal would be seriously detrimental to the history, architecture and C
> visual character of this part of the Cartmel Conservation Area and would be contrary to policies C5 and C23 of the Cartmel and Furness Local Plan and C14/C15 of the Cumbria and Lake District Joint Structure Plan, First Alteration.

On March 9 1989 the Parsonages Board appealed to the Secretary of State against the refusal of permission. The Secretary of State appointed an D
inspector (Mr N A C Holt RIBA) to determine the appeal. The appeal was allowed on July 13 1989 upon a consideration of written representations and after a site inspection. On August 21 1989 the district council applied for the decision to be quashed. The application was successful and hence the present appeal. The Parsonages Board, although formally the second respondents, have not appeared either in this court or below.

The argument before this court focused, as it did in the court below, upon E
section 277(8) of the Act of 1971. It provided, so far as is material:

> Where any area is for the time being designated as a conservation area, special attention shall be paid to the desirability of preserving or enhancing its character or appearance in the exercise, with respect to any buildings or other land in that area, of any powers under this Act, . . .

The determination of an appeal is an exercise of power under the Act (see F
section 30(3)). There is no dispute but that the inspector did appreciate his need to pay special attention to the desirability of preserving or enhancing the character or appearance of the Cartmel conservation area. He expressed his appreciation in para 2 of the decision letter in these words:

> I am of the opinion that the main issues are firstly, what effect the proposal
> would have on the character and appearance of the Cartmel Conservation G
> Area, having regard to the desirability of preserving or enhancing that character and secondly, . . .

The argument accepted by the learned deputy judge was that the inspector had failed to discharge the duty which he had recognised in that he had not addressed the question of whether the proposed new vicarage would, or would
not, preserve the character or appearance of the Cartmel conservation area. H
The material paragraphs of the decision letter are these:

> 6. The Vicarage in my opinion should be regarded as being within the confines of the village. Whilst the proposed development would not fall within the generally accepted definition of infilling, I do not consider that it would be contrary to the important objectives of the settlement policy, designed to protect the countryside from unnecessary development. Providing that the proposed house did not cause harm to the character of the

> A Conservation Area, I consider that it would accord with Policy A3 of the Plan. As regards Policy C5, in my opinion the local authority are fully justified in protecting the open areas within the village, which make a significant contribution to its character. I do not consider however that the mature domestic curtilage of the Vicarage, which is to a large extent screened from public vantage points by trees and shrubs along the east and west boundaries, and by the stone wall along the frontage to Priest Lane, should B be seen in the same light as the nearby open pasture land.
>
> 7. I would accept that the proposed house would be visible from Priest Lane, over the existing wall in front of the site and also when approaching from the east. I consider however that because of the wall and the existing trees and shrubs on the site, that would be retained, the impact of a new house would not be great.
>
> In my opinion the effect on the character and appearance of this part of the C Conservation Area would be small. I am also satisfied that the grounds of the Vicarage are sufficiently large to accommodate a new house without serious detriment to the setting of the existing building and without affecting the larger trees within the garden that make a particularly significant contribution to the area.
>
> 8. I very much appreciate the concern of the Council and the local people, to preserve and enhance the special quality of Cartmel, and I would agree that they should be strongly supported. I am of the opinion however that D providing great care was exercised in the detailed design of the proposed house, having regard to the precise siting, the materials, the massing, the roof pitch, and the details of features such as the eaves and windows, the proposed Vicarage could be accommodated without damaging consequences to the appearance of the village. Whilst there might have been no new building of significance in this part of Cartmel for over 100 years, that is not a reason to prevent development now, if no harm would result.
>
> E 9. I have had regard to the potential precedent that could be established were this appeal allowed. In the context of the village, I am satisfied that the physical characteristics of the site make this a special case and I do not consider that a permission for your client's proposal would make it difficult for the Council to refuse applications elsewhere that might have more damaging consequences to the character of what is undoubtedly a most important conservation area.

F These paragraphs were examined before us in order to show that the inspector either had or had not performed his duty under section 277(8). How, then, should he have approached the performance of his duty? There was no agreement upon the answer to this question, which is a question of general importance.

The debate was to whether, in paying special attention to the desirability of G preserving or enhancing the character or appearance of a conservation area, the decision-taker is bound to determine whether the proposed development would make a positive contribution to preservation and enhancement. If he is, then a determination that neither character nor appearance is harmed is an insufficient discharge of the duty, for the decision-taker will not have answered the correct question. I should say that whatever be the correct question, its answer is not necessarily decisive of a development proposal, for H there may in particular circumstances be other material considerations which are favourable to, or adverse to, a proposal which fails or passes the conservation test. However, in practice the conservation consideration will be a weighty factor in the mind of a decision-taker.

The learned deputy judge answered the question in accordance with his own previous decision in *Steinberg* v *Secretary of State for the Environment* [1989] 2 PLR 9. In that case he said (at p 12F):

> There is, in my judgment, a world of difference between the issue which the inspector defined for himself — whether the proposed development would "harm" the character of the conservation area — and the need to pay special attention to the desirability of preserving or enhancing the character or appearance of the conservation area. In short, harm is one thing; preservation or enhancement is another. No doubt the inspector has demonstrated his concern that the character of the conservation area should not be harmed. That, in my judgment, is not the same as paying special attention to the desirability of preserving or enhancing that character as well as its appearance. The concept of avoiding harm is essentially negative. The underlying purpose of section 277(8) seems to me to be essentially positive.

In *Harrow London Borough Council* v *Secretary of State for the Environment* (1989) 60 P&CR 525 Sir Graham Eyre QC, sitting as a deputy judge of the Queen's Bench Division, seems to have thought in different terms. He said (at p 529)*:

> As I understand the subsection, it does not set out the test nor does it support the proposition that in relation to a consideration of development proposals they must themselves preserve or enhance, or serve to preserve or enhance, the character or appearance of a conservation area. It imposes a duty to pay special attention to the desirability of preservation or enhancement.

In seeking to resolve the issue I start with the obvious. First, that which is desirable is the preservation or enhancement of the character or appearance of the conservation area. Second, the statute does not in terms require that a development must perform a preserving or enhancing function. Such a requirement would have been a stringent one which many an inoffensive proposal would have been inherently incapable of satisfying. I turn to the words. Neither "preserving" nor "enhancing" is used in any meaning other than its ordinary English meaning. The court is not here concerned with enhancement, but the ordinary meaning of "preserve" as a transitive verb is "to keep safe from harm or injury; to keep in safety, save, take care of, guard" (*Oxford English Dictionary* (2nd ed) vol XII, p 404). In my judgment, character or appearance can be said to be preserved where they are not harmed. Cases may be envisaged where development would itself make a positive contribution to preservation of character or appearance. A work of reinstatement might be such. The Parsonage Board never advocated the new vicarage on that basis. It was not a basis which the inspector was invited to address but, importantly, he did not have to address it because the statute does not require him so to do.

The statutorily desirable object of preserving the character or appearance of an area is achieved either by a positive contribution to preservation or by development which leaves character or appearance unharmed, that is to say, preserved. Development of the latter type was said by Mr John Laws, for the Secretary of State, to be of a neutral character and therefore unobjectionable by reference to the conservation consideration. It was, said Mr Laws, development to be found in the unassigned middle ground between the poles occupied by development which of itself preserved character and development which harmed character. Where development is neither positive nor neutral in regard to preservation it must inevitably be development which would to some degree harm character or appearance, and the decision-taker will have to consider whether to grant permission notwithstanding that harm. If he decided to do so, he would have to have a reason for setting aside the enjoined conservation consideration. Such a reason might be, for example, that the

*Editor's note: Also reported at [1990] 2 PLR 62 at p 66C.

A alteration must be accepted so that a development desirable in some public interest other than that in conservation can take place.

It follows from what I have said that I respectfully disagree with the view of the learned deputy judge expressed in *Steinberg* and which he applied in the present case.

Our attention was drawn to the recent decision of another constitution of this court on February 6 1991 in *Bath Society* v *Secretary of State for the* B *Environment and Hammercrest Developments Ltd**. The leading judgment was given by Glidewell LJ with whom Stocker LJ and Sir Denys Buckley agreed. In that case the Secretary of State had accepted his inspector's recommendation that an appeal against a decision refusing permission for a development in a conservation area should be allowed. The Secretary of State's decision was challenged on the ground that he had failed to comply C with the duty imposed on him under section 277(8). The Secretary of State had accepted the inspector's conclusions so that in effect the challenge was to the inspector's reasoning. The challenge failed before Hutchison J but succeeded in this court.

In that case the inspector concluded that the impact of the proposed development upon a view within the conservation area "would not be unacceptably great". There was implicit in that conclusion a finding that there D would be harm. The inspector had not found a neutral position, so that the court was not concerned with such a position. Having implicitly found that there would be harm, the inspector should have addressed the question of whether there was a reason for setting aside the enjoined conservation consideration. This he did not do, and his failure was the basis of Glidewell LJ's judgment. In the course of his judgment, Glidewell LJ gave guidance upon the approach which should be adopted when an application is made for E permission to develop in a conservation area and at p 25 of the handed-down judgment† he said:

> If. . . the decision-maker decides that the proposed development will neither preserve nor enhance the character or appearance of the conservation area, then it is almost inevitable that the development will have some detrimental, ie harmful, effect on that character or appearance.

F Mr Nigel Macleod QC, for the first respondent, relied upon this passage as showing that a proposal must have a preservative function before it satisfies the conservation consideration. I do not think that the passage sustains this proposition. The inevitability of harm cannot have been supposed by Glidewell LJ to be a consequence which peculiarly follows an absence of positive contribution, for it follows also where the development is not neutral. Mr Macleod's approach would have Glidewell LJ guilty of an offence against G logic by propounding a *non sequitur*. Mr Macleod also drew our attention to a passage at p 28 of the handed-down judgment§ where, in dealing with a contention on behalf of the Secretary of State, Glidewell LJ said:

> Mr Howell argues that a conclusion that a proposed development will do no harm is equivalent to a conclusion that it will preserve. Even if that is correct (and adopting the approach of Mr Lionel Read QC I doubt whether it is) this is not what in my view the Inspector himself concluded.

H

I also, and with respect, think that an absence of harm affords no basis for concluding that there would be a positive contribution to conservation. This is, however, beside the point in the present case.

I have dealt with the *Bath* case at some length because it is a recent decision

*Editor's note: Reported at p 51 *ante*.
†Editor's note: Reported at p 65C *ante*.
§Editor's note: Reported at p 66D *ante*.

of this court upon which reliance was placed. However, I do not find in it anything inconsistent with what I have said in this judgment. A

I turn to whether the inspector in the instant case did properly do his duty. The case before him was one where neither positive preservation nor enhancement was asserted. The inspector looked to harm. He looked for the neutral position.

In examining how he expressed the performance of his task in the paragraphs which I have quoted, it is to be remembered that the examiner is considering a letter written for an informed audience and which is not to be construed as if it were a contract or an enactment. In my judgment, it can fairly be said that the inspector found that neither the character nor the appearance of the Cartmel conservation area would be harmed by the new vicarage and that its character and appearance would remain preserved notwithstanding the development. The important paragraphs are those numbered 7 and 8. In para 7 I draw attention to "character and appearance" and "would be small". In para 8 I draw attention to the opening sentence and to "without damaging consequences". The word "if" in the final sentence can sensibly be read only as "if as here". I confess to at one time feeling anxiety about the phrase "more damaging" in para 9, which seems to suppose that the proposal would have damaging consequences. I am now persuaded that to make that supposition is to make nonsense of what has gone before. That should not be done. All is sense if the word "more" is ignored. I suspect that the word, which I have found in many decision letters, is incantatory, and is to ward off accusations that a dangerous precedent is being established. B C D

For the reasons which I have given, I would allow this appeal and set aside the order of Mr Lionel Read QC.

SIR CHRISTOPHER SLADE: I gratefully adopt Mann LJ's statement of the facts of this case and I agree with his conclusion. E

Para 15 of Planning Policy Guidance Note 1 — *General Policy and Principles* issued by the Department of the Environment ("PPG1") states:

> 15 . . . There is always a presumption in favour of allowing applications for development, having regard to all material considerations, unless the development would cause demonstrable harm to interests of acknowledged importance, . . . F

In *Bath Society* v *Secretary of State for the Environment and Hammercrest Developments Ltd* Glidewell LJ, with whose judgment Stocker LJ and Sir Denys Buckley agreed, helpfully stated six principles which should govern the approach to an application for planning permission for development in a conservation area, having regard to provisions both of section 277(8) of the Town and Country Planning Act 1971 and of PPG1. After setting out five principles, with which I am in entire agreement, Glidewell LJ stated*: G

> (vi) If, however, the decision-maker decides that the proposed development will neither preserve nor enhance the character or appearance of the conservation area, then it is almost inevitable that the development will have some detrimental, ie harmful, effect on that character or appearance. In lawyers' terms, the presumption derived from the policy in para 15 is then rebutted. As I have said, the conclusion that the development will neither enhance nor preserve will be a consideration of considerable importance and weight. This does not necessarily mean that the application for permission must be refused, but it does, in my view, mean that the development should only be permitted if the decision-maker concludes that it carries some advantage or benefit which outweighs the failure to satisfy the section 277(8) test and such detriment as may inevitably follow from that. H

*Editor's note: Reported at p 65C *ante*.

This sixth principle is, it seems to me, with respect, no less entirely correctly stated, provided only that the word "preserve" is there intended and read in a sense corresponding with the word "preserving" in section 277(8). Perhaps the greater part of the argument before this court has, however, centred on the question whether the word preserving in section 277(8) is apt to include only development which will make a positive contribution to the character or appearance of the conservation area in question (the narrower sense) or is apt to include, in addition, development which, while making no such positive contribution, will do no harm to such character or appearance (the wider sense). This was a question which this court did not have to decide in the *Bath Society* case. Glidewell LJ touched on it in the two sentences quoted by Mann LJ in his judgment, where he appears to have inclined to favour the narrower construction of the word "preserving" adopted by Mr Lionel Read QC in *Steinberg* v *Secretary of State for the Environment* [1989] 2 PLR 9. In that case the deputy judge had said (at p 12G):

> The concept of avoiding harm is essentially negative. The underlying purpose of section 277(8) seems to me to be essentially positive.

The word "preservation" in section 277(8) is intended to include only positive preservation.

Glidewell LJ, however, did not decide the point and his provisional view was only tentatively expressed in the passing comment at p 28 of his judgment*. I venture to think that if he had had the benefit of full argument on the point such as we have had in this court, it might well have been somewhat differently expressed. This inference is reinforced by the consideration that if in using the word "preserve" in para 6 of his judgment, quoted above, he had intended to refer solely to development which made a positive contribution to preservation and to exclude development of a neutral character, the statement beginning with the words "then it is almost inevitable . . ." would have been difficult to follow.

In the absence of any authority binding this court as to the meaning of the word "preserving" in section 277(8) of the Act of 1971, we are free to decide the point. For the reasons given by Mann LJ, to which I wish to add only very little, I am of the clear opinion that the wider sense of the word is the correct one.

The word "preserving" is used in the subsection in conjunction but in contrast with the word "enhancing", which itself imports the notion of a positive improvement. In my judgment, on the ordinary meaning of words, the character or appearance of an area can properly be said to be preserved by a development where they are not harmed, even though the word would be no less apt to include preservation by restoration or reinstatement. The statutory justification for designating an area as a conservation area is itself expressed by reference to the designator's determination that it is "desirable to preserve or enhance" the character or appearance of the area in question (see sections 277(1), (2) and (4)). I cannot think that in those three subsections the legislature intended the word "preserve" to bear the narrower rather than the wider sense.

Any exercise of the power to grant planning permission is likely to involve at least some *alteration* of the *appearance* of the buildings or other land in question. Furthermore, the draftsman of section 277, instead of using the phrase "character and appearance", has throughout the section chosen to follow the references in the alternative to preservation or enhancement with references in the alternative to character *or* appearance. With these two points

*Editor's note: Reported at p 66 *ante*.

in mind, after a first introduction to the section, I tentatively questioned A whether it might not possibly be right in subsection (8) to read the word "preserving" in its context as intended to govern the word "appearance". However, I understand from counsel well versed in these matters that this suggestion was a novel one, which so far as they knew derives no support whatever from any source. I therefore do not wish to pursue this particular heresy. B

This being so, section 277(8) obliges the decision-maker who is dealing with an application for planning permission for the development of land in a conservation area to give special attention to two questions:

(1) Would the proposed development preserve the character or appearance of the conservation area?

(2) Would it enhance the character or appearance of that area? C

As regards the second of these two questions, I agree with the deputy judge that it is implicit in the inspector's decision in the present case that he did not think that the proposed development would *enhance* the character or appearance of the Cartmel conservation area.

As regards the first question, the inspector did not in terms give an affirmative answer to it. However, respectfully disagreeing with the deputy judge on this point, I think that the inspector did give it in the affirmative by D clear and necessary inference. In para 2 of his decision letter, he had specifically identified the first of the main issues as being "what effect the proposal would have on the character and appearance of the Cartmel conservation area, having regard to the desirability of preserving or enhancing that character". It is not disputed that this main issue was correctly, if somewhat loosely, identified and that the inspector, in stating it, clearly had the provisions of section 277(8) in mind. (The case is in this respect E distinguishable from *Steinberg*'s case, where the inspector had not mentioned the subsection or his duty thereunder either in terms or by the use of any language from which it might reasonably be inferred that he was intending to refer to it: see [1989] 2 PLR 9 at p 11C-D.)

Once it is recognised that the inspector had correctly identified his statutory duty, the court should, in my judgment, be slow to infer from the subsequent F passages in his decision letter which deal with the conservation issue that he was not thereby intending to discharge that duty.

Subject to what will be said in the next two paragraphs of this judgment, I feel little doubt that in paras 7 and 8 of his decision letter (which have already been quoted by Mann LJ in his judgment) the inspector was intending to discharge and did discharge the duty imposed on him by section 277(8). In my judgment, these paras 7 and 8, read fairly in their context, following the G identification of the first main issue in para 2, should reasonably be understood as expressing the inspector's conclusion that, while the proposed development would not enhance the character or appearance of the conservation area, its effect on such character or appearance would be small and it *would* preserve such character or appearance provided only that "great care was exercised in the detailed design of the proposed house, having regard H to the precise siting, the materials, the massing, the roof pitch and the details of features such as the eaves and windows". No doubt so as to ensure that such care would be exercised, the inspector (by para 13) granted planning permission subject to the condition, among others, that "approval of the details of the siting, design and external appearance of the building, the means of access thereto and the landscaping of the site . . . shall be obtained from the local planning authority".

I think it necessary to deal only with two further points. The inspector, in para 7 of his decision, stated:

> I am also satisfied that the grounds of the Vicarage are sufficiently large to accommodate a new house without serious detriment to the setting of the existing building.

This passage, in my judgment, gives rise to no difficulties. It is the effect on the character or appearance of the conservation area as a whole to which section 277(8) requires attention to be directed — not the effect on the character or appearance of particular buildings within that area.

Second, the inspector (in para 9, which has also been quoted in Mann LJ's judgment) used the phrase "applications elsewhere that might have more damaging consequences to the character of what is undoubtedly a most important conservation area". Para 9, when read as a whole, is clearly intended as no more than a warning to future applicants for planning permission for development in the Cartmel conservation area that the inspector regarded the application before him as a special case and that it should not be regarded as a precedent. In my judgment, the use by the inspector of the phrase "more damaging" in para 9 cannot fairly be read as derogating in any way from his conclusion expressed in paras 7 and 8 that the effect of this particular development on the character or appearance of this part of the conservation area would be small and no harm would result from it.

In my judgment, therefore, this is not a case falling within principle (vi) stated by Glidewell LJ in the *Bath Society* case. In my judgment, on a fair reading of paras 7 and 8 of his decision letter, the inspector did decide that the proposed development would preserve the character or appearance of the conservation area. He was entitled so to decide and, in so doing, did not misdirect himself as to the nature of the duty imposed on him by section 277(8). In my judgment, he duly discharged that duty.

I, too, would allow this appeal.

BUTLER-SLOSS LJ: I agree with the judgment of Mann LJ and that of Sir Christopher Slade, both of which I have had the opportunity of reading in draft, and only add some brief comments since we are differing from a most experienced deputy High Court judge.

The problem arises as to the meaning of the word "preserving" in section 277(8) of the 1971 Act and whether the proposed development must positively preserve the character or appearance of the conservation area in question. In this case no suggestion was ever made that the proposed new vicarage either enhanced or positively preserved the character or appearance of the village. The most that could be said was, as was found by the inspector, that "it could be accommodated without damaging consequences to the appearance of the village". The development would cause no harm and its impact was consequently neutral. I agree with Mann LJ for the reasons which he gave that no weight should be placed upon the phrase "more damaging" in the final paragraph of the inspector's decision.

What is the correct approach under section 277(8) to a situation where the proposed development is neutral? The starting point is the normal meaning of the word "preserve", which includes to keep safe from harm, to take care of or guard. Consequently in ordinary language, that which is kept safe from harm is preserved. In my judgment, "preserve" in the section is to be looked at in the same way. Although some development may indeed be seen as positively preserving the character or appearance of an area, I do not consider that the Act requires such a limitation on the construction of the subsection. I find support for that view in section 277(1), which states:

> Every local planning authority shall from time to time determine which parts of their area are areas of special architectural or historic interest the character or appearance of which it is desirable to preserve or enhance, and shall designate such areas as conservation areas. A

A construction of preserve to require a positive approach might well be incompatible with the application of this subsection. Further, if positive preservation is the correct construction, what is to be done with the middle, neutral ground which, as this case demonstrates, does exist? In the decision of B
Mr Lionel Read QC in *Steinberg* v *Secretary of State for the Environment* [1989] 2 PLR 9 he appeared to be assuming that unless a development was positively preserving it was to be considered as negative and consequently did not fall within section 277(8). Not only does the subsection not specifically express that requirement but it would seem to me to be at variance with the general presumption in favour of allowing applications for development, C
having regard to all material considerations, unless the development would cause demonstrable harm to interests of acknowledged importance: see PPG1.

Glidewell LJ in *Bath Society* v *Secretary of State for the Environment and Hammercrest Developments Ltd* set out six propositions of principle to be derived from the Act and the decided cases with which, with one qualification, I respectfully agree. Proposition (vi) states: D

> If, however, the decision-maker decides that the proposed development will neither preserve nor enhance the character or appearance of the conservation area, then it is almost inevitable that the development will have some detrimental, ie harmful, effect on that character or appearance.

To read the sentence with the restrictive meaning of "preserve" does not, in my view, make entire sense and may be said to be a *non sequitur*. Glidewell LJ E
was not considering the neutral position in the *Bath* case and cannot have had it in mind in this passage. I do not agree with Mr Macleod's argument in favour of the narrower interpretation, nor do I see Glidewell LJ as supporting it despite a passage later in his judgment, to which Mann LJ has referred.

In my judgment, the inspector, and consequently the Secretary of State, paid proper regard to the requirements of section 277(8). I agree that this appeal should be allowed and the order of Mr Lionel Read QC be set aside. F

Appeal allowed with costs.

Richmond upon Thames London Borough Council *v* Secretary of State for the Environment and another G

QUEEN'S BENCH DIVISION
Henry J
March 12 1991

Parapet walls erected on flat roof — Whether alteration of roof — Whether enlargement of dwelling — Whether permitted development H

In the mid 1970s an extension with a flat roof was constructed at the second respondent's house at 59 St Margaret's Grove, Twickenham, Middlesex. The second respondent then erected parapet walls around the flat roof. The appellant council issued an enforcement notice on June 15 1989 alleging that a breach of planning control had occurred by the erection of the walls. The

A second respondent appealed, though his notice of appeal did not refer to ground (*b*), of section 88(2) of the Town and Country Planning Act 1971. The appeal was allowed by the Secretary of State for the Environment by his inspector, who concluded that the building work carried out was permitted development within the General Development Order 1988, Schedule 2, Part 1, Class B, being an enlargement of a dwellinghouse.

The council appealed that decision under section 289 of the Town and
B Country Planning Act 1990 contending that the inspector erred in deciding that the works were an alteration or addition to the roof or were an enlargement within the GDO; the roof was not altered and the erection of the walls did not cause an "enlargement" of the dwellinghouse within the meaning of Class B because there was no increase in its volume.

Held The appeal was dismissed.

The inspector concluded that the addition of the walls, together with the
C previous extension, did enlarge the dwellinghouse; but that enlargement, on a balance of probabilities, did not exceed the limits set out in Class B1 (*d*) of the GDO. In deciding whether the work was an alteration or an enlargement, the correct approach must be to proceed on the external appearance of the dwellinghouse. The inspector was entitled to find the construction of the walls in themselves enlarged the dwelling and that their erection was permitted development within Class B1: see p 111.

D

Case referred to in the judgment

Bradford Metropolitan District Council v *Secretary of State for the Environment* (1977) 35 P&CR 387; 245 EG 575; [1978] JPL 177

Appeal under section 289 of the Town and Country Planning Act 1990

This was an appeal under section 289 of the Town and Country Planning Act 1990 against the decision of the first respondent, the Secretary of State for the
E Environment, by his inspector, who had allowed an appeal by the second respondent, Mr Neale, who appealed against an enforcement notice issued by the appellant local planning authority.

Anne Williams (instructed by the solicitor to Richmond London Borough Council) appeared for the appellants.

David Holgate (instructed by the Treasury Solicitor) appeared for the first respondent, the Secretary of State for the Environment.

F *The second respondent, Mr Neale, did not appear and was not represented.*

The following judgment was delivered.

HENRY J: On June 15 1989 Richmond upon Thames London Borough Council served an enforcement notice on Mr Neale, the occupier of a house in Twickenham, at 59 St Margaret's Grove. The works described in that enforcement notice were the erection of a parapet wall on the roof of a first-
G floor extension to that house. The works are described in greater detail in the decision letter following the inquiry where the inspector writes as follows:

> The appeal premises consist of a two storey mid-terrace house on the north side of St Margaret's [Grove] in an area which is predominantly residential. As built, the dwelling in common with other nearby dwellings extended to the rear incorporating two storeys and a north facing gable end which also encompassed No 57 to the east.
>
> H The premises have been extended and adapted, notably by the recent addition of a dormer extension filling in the rear roof slope and by earlier works which created a flat roof on the rear facing two storeys. Access to the flat roof is available via a narrow window approximately 0.52m wide and 1.75m high hinged at the top and swinging open vertically. The window has limited opening and the access is extremely awkward, partly relying on movable steps on the roof. The flat roof has a parapet boundary on the east side formed by the original division wall between the terraced properties.

> There is a former chimney at the north east corner and the south (*sic*) and west sides of the flat roof have a .255m thick brickwork parapets approximately one metre high, similar in height to the pre-existing east wall. Although not specifically referred to in the notice it is the parapet walls on the north and west sides only which are the subject of the notice.

On June 20 1989 the householder appealed against that enforcement notice. He referred to the statutory grounds in his notice of appeal but made no reference to ground (*b*), no breach of planning control, though that was the predominant issue at the inquiry when it took place.

The hearing was on January 16 1990 and by his decision letter of February 19 1990 the inspector made the crucial finding in this case, namely:

> I consider that the development enforced against can properly be described as "The enlargement of a dwellinghouse consisting of an addition or alteration to its roof", therefore falling to be assessed against the criteria of Schedule 2 Part 1 Class B of the 1988 GDO.

He went on to find that the development did not infringe the conditions of Class B subject to which development is permitted. He therefore found that the development — the erection of the parapet walls — did not constitute a breach of planning control because planning permission had been given to them by virtue of the Town and Country Planning General Development Order 1988 (SI 1988 No 1813) Class B, the permitted development being the enlargement of a dwellinghouse consisting of an additional alteration to its roof.

It is that finding which is the subject of challenge by the council in these proceedings before me. They contend that the inspector erred in law in that it was not open to him in law to find that the erection of the parapet wall on the north and the west side of the premises was an enlargement of the dwellinghouse. The contention is that to effect an enlargement of the dwellinghouse one must have an increase in the volume or usable space within the dwellinghouse and, of course, the erection of a wall on a flat roof would not achieve that.

It is necessary, in looking at this matter, to look at the scheme of the general development order. The 1988 general development order replaced the 1977 order with certain changes that are relevant to this case. In that order one gets classes of permitted development from A-H. After each definition of the permitted development there is a list of exceptions, "Development not permitted". The starting point is the permitted development in Class A, namely, "The enlargement, improvement or other alteration of a dwellinghouse". Those words are the same, effectively, as in Class A in the 1977 order, though the list of development not permitted which follows is significantly different.

Looking first at those words as they stand, clearly the parapet walls, at lowest, would be an alteration of the dwellinghouse. One sees the court reaching a conclusion similar to that in relation to Class A in the case of *Bradford Metropolitan District Council* v *Secretary of State for the Environment* (1977) 35 P&CR 387, which is a decision of Griffiths J on October 14 1977. I read the headnote of that case:

> The second and third respondents wished to face their red-brick houses with stone tiles, and they applied to the local planning authority for planning permission for that purpose. In each case, the stone cladding would be of a total thickness of one-and-a-quarter inches and would therefore project that distance beyond the original brickwork, but it would be within the building line if the building line were taken to be the edges of the sills and the moulded surrounds to the lower windows. The local planning authority refused

> permission on aesthetic grounds. On appeal by the second and third respondents, the Secretary of State expressed the view that on the facts of the cases the sills and moulded surrounds should be regarded as part of the wall and that as the proposed cladding would not project beyond them it fell within the development permitted by article 3 of and Class I.1 of Schedule 1 to the Town and Country Planning General Development Order 1977 and planning permission was not required for it. He accordingly took no further action on the appeals. The local planning authority applied under section 245 of the Town and Country Planning Act 1971 for the decision of the Secretary of State to take no further action to be quashed.
>
> **Held**, dismissing the appeals, (1) that the development permitted by paragraph (*c*) of Class I.1 of Schedule 1 of the Order of 1977 was not limited to structural as opposed to cosmetic features; that those were, in any event, not mutually exclusive concepts; and that the words of Class I.1 were wide enough to cover such an operation as stone cladding.

In the course of his judgment Griffiths J, as he then was, said this of the stone cladding at p 389:

> I now turn to Schedule 1. Schedule 1 is divided into various classes, and the two classes with which I shall be concerned are Class I and Class II. Class I provides: "Development within the curtilage of a dwelling house. 1. The enlargement improvement or other alteration of a dwelling house so long as: . . .," and there follow a number of provisions. Giving these words their ordinary meaning to add stone cladding to the face of a building is clearly at least an alteration of the dwelling-house. Presumably in the mind at least of the occupier it is an improvement of the dwelling-house, and perhaps it is arguable that it is an enlargement, only that I doubt.

It is not clear from that passage, where Griffiths J has to deal with whether the addition of the stone cladding is an enlargement of the dwellinghouse, whether he is doubting that it is because it would be *de minimis* to the dwellinghouse or because, as the appellants would contend, stone cladding adds no extra usable space for the householder in the dwellinghouse.

As I have said, clearly the parapet wall is an alteration to the roof in the ordinary meaning of that term analogous to that extent, at any rate, with the cladding. However, such an alteration is not permitted by Class A because under A.1 "Development not permitted", (h) one has the wording, "it would consist of or include an alteration to any part of the roof", which this clearly does.

If it is an alteration to any part of the roof, the regime dealing with it is either Class B or Class C and they are alternatives as will be seen. Class B deals with the enlargement of a dwellinghouse consisting of an addition or alteration to its roof, and Class C deals with any other alteration to the roof of a dwellinghouse other than the enlargement of a dwellinghouse consisting of an addition or alteration to its roof. Those different classes are mutually exclusive and, where one has an alteration to any part of the roof, for the development to be permitted it must come within one or the other and, on the wording of the order, it could not come within both.

If the permitted development is an enlargement, as the inspector here found, it must not fall foul of the conditions set out in B.1 (a)-(e), which deal with the size, the height and the extension of the enlargement beyond the plane of the roof space fronting the highway. If, on the other hand, the development is not an enlargement but is any other alteration to the roof of a dwellinghouse, there is a single condition that must be complied with. Development not permitted is, in those circumstances, a development which would result in a material alteration to the shape of the dwellinghouse.

The problem therefore is clearly exposed. Here two of the flank walls of the back addition have been raised. Externally the building has been enlarged because the parapet raises the level of the aspect of the north and west sides and so, to that extent, makes the house apparently larger, but there is no more space within the dwellinghouse and the question is whether the word "enlargement" in the general development order is to be restricted to such enlargements which give more useful enclosed accommodation to the occupier, from whose point of view the erection of the parapet wall on the roof would not enlarge his dwellinghouse at all: ie does enlargement mean enlarging the dwelling-space in the dwellinghouse, the enclosed space, or does it mean the house apparently appearing larger to those outside it looking at it? This is a matter of pure statutory interpretation of the natural ordinary meaning of language in the context of the general development order.

First, to clear the point out of the way, I get no help from the fact that the order deals with buildings that are dwellinghouses. That simply defines the buildings that the order applies to. It does not necessarily follow from the use of the word dwellinghouse that the only form of enlargement of such a building is an enlargement to the usable dwelling-space enclosed within the house, the volume of enclosed space within the four walls and the roof of that house.

Clearly, as I have said, the continuation of the existing east parapet wall round the other two open sides of the flat roof of the rear, the north and the west, is undoubtedly an alteration to the roof of the dwellinghouse. The question whether the erection of those parapet walls forms an enlargement of the dwellinghouse depends on whether you have regard only to the external appearance of the house or whether you judge it by the amount of internal enclosed space. From its external appearance the building is undoubtedly enlarged because its north and west flank walls are higher, but there is no more usable space within the house. The householder would not regard his dwelling as enlarged. His neighbour, on the other hand, would regard the building as enlarged.

It seems to me, in all the circumstances, that the correct approach must be to proceed on the external appearance of the house. First, planning restrictions are to do with external appearances, on what is apparent to the neighbourhood and not on internal practicalities or realities. Second, some support for that approach is to be found in that, in calculating whether an enlargement of the building is too large to be permitted by the restrictions which impose limits on the cubic increase on the size of the building, the general development order provides that the cubic content of the building before, and the building after for comparison purposes, is to be measured externally and not internally. In these circumstances I conclude that the inspector was entitled to find, as a matter of law, that the development in question constituted an enlargement of the dwellinghouse. Therefore, this appeal fails.

Appeal dismissed with costs.

Ramsey and another *v* Secretary of State for the Environment and another

QUEEN'S BENCH DIVISION

Mr Gerald Moriarty QC (sitting as a deputy judge of the Queen's Bench Division)

March 14 1991

A **Enforcement notices — Planning unit — Two notices directed to same activity but relating to overlapping areas — Whether lawful to issue two enforcement notices**

The first appellant and his brother farm 112 ha at Hill Farm, Rushmere St Andrew, Ipswich. They are the directors of the second appellant company, Ramsey Sports Ltd. The company operates on a 7.6 ha site, part of the 112 ha,
B a motor park for vehicular sports and leisure. The second respondents, Suffolk Coastal District Council, issued two enforcement notices, each requiring the use for sports and leisure activities to cease. One notice related to a smaller site where "motocross" activity took place and which was occupied by the second appellants; the second notice related to a larger area of land being 112 ha in the occupation of the first appellant. The appellants contended that it was not lawful for the respondent council to serve, and for the Secretary
C of State to uphold, two enforcement notices in relation to two sites, one of which overlaps or includes the other.

Held The appeal was dismissed.

There is no requirement in sections 172 and 173 of the Town and Country Planning Act 1990 to limit the enforcement procedure to a site or sites which do not overlap one another. There is no requirement of exclusivity in sections
D 174 and 175 in relation to the appeal stage. There is no indication that several enforcement notices cannot properly be served in relation to the same site, or in relation to one activity on a site that is defined independently of any other site, or that is defined in such a way as to form part of or be overlapped by or overlap a second site: see p 118H.

There are good practical reasons not to impose any strict requirements which are not expressly derived from the statute. There are no constraints imposed by the decided cases. The enforcement notice procedure should not
E be used even inadvertently to the effect that a landowner's rights are cut down, because it is designed only to be utilised against development which is in breach of control: see p 119A. The effect of upholding either of the two enforcement notices is not to restrict the rights of the landowner. If any planning authority were to seek to prosecute under both notices in respect of the same alleged activity, that would be unjustified and it would not be accepted by the courts: see p 119F.

F

Cases referred to in the judgment

Burdle v *Secretary of State for the Environment* [1972] 1 WLR 1207; [1972] 3 All ER 240; (1972) 70 LGR 511; 24 P&CR 174; [1972] EGD 678; 223 EG 1597, DC

de Mulder v *Secretary of State for the Environment* [1974] QB 792; [1974] 2 WLR 533; [1974] 1 All ER 776; (1973) 72 LGR 371; 27 P&CR 369, DC

Johnston v *Secretary of State for the Environment* (1974) 28 P&CR 424, CA

Rawlins v *Secretary of State for the Environment* [1989] JPL 439

G *Trentham (G Percy) Ltd* v *Gloucestershire County Council* [1966] 1 WLR 506; [1966] 1 All ER 701; (1966) 64 LGR 134; 18 P&CR 225, CA

Appeal under section 289 of the Town and Country Planning Act 1990

This was an appeal under section 289 of the Town and Country Planning Act 1990 against the decision of the Secretary of State for the Environment by his inspector,

who had dismissed appeals by the appellants against two enforcement notices A
issued by the second respondents, Suffolk Coastal District Council.

Michael Bedford (instructed by Howes Percival, of Ipswich) appeared for the appellants, Reginald William Ramsey and Ramsey Sports Ltd.
Michael Kent (instructed by the Treasury Solicitor) appeared for the first respondent, the Secretary of State for the Environment.
The second respondents, Suffolk Coastal District Council, did not appear and B
were not represented. Cur adv vult

The following judgment was delivered.

MR GERALD MORIARTY QC: This is an appeal under section 289 of the Town and Country Planning Act 1990 against the decision of the Secretary of State for the Environment by his inspector dated September 27 1990. The inspector dismissed appeals under section 174 of the 1990 Act against two C
enforcement notices referred to in the course of the case as notice 1 and notice 3. The first appeal was on grounds (a) and (h) and the second on grounds (a), (c) and (h) of section 174(2) of the 1990 Act.

The appeal to this court was on the two grounds stated in the notice of motion; first, error of law by the Secretary of State in dismissing the appeal against enforcement notice 3 on ground (c) and, second, error in law in upholding both enforcement notice 1 and enforcement notice 3. However, on D
the appeal Mr Bedford, on behalf of the appellants, did not proceed with the first ground and this appeal, therefore, raises a single issue, in effect whether it is lawful to serve and uphold two enforcement notices in relation to two sites, one of which overlaps or includes the other, in order to secure the cessation of a single activity which is carried out exclusively on the smaller of the two sites. It is relevant to note at once that there was no appeal to the Secretary of State under section 174(2)(g), that is that the steps required by either notice exceeded E
what was necessary to remedy any breach of control or to achieve a purpose specified in section 173(4).

The appellant, Reginald William Ramsey, and his brother farm 112 ha at Hill Farm, Rushmere St Andrew, Ipswich. In 1987 they formed and became directors of the appellant company. The company operates a site of 7.6 ha, part of the 112 ha of the farm holding. This site is called Suffolk Motopark. F
They operate it for vehicular sports and leisure. That comprehends use by a variety of vehicles, including scramble bikes and four-wheel-drive vehicles.

The first requirement of enforcement notice 1, and the first requirement of enforcement notice 3, was in each case to cease the use for sports and leisure activities.

The inspector dealt with the validity of enforcement notice 3 in relation to the larger site on the issues raised by the appeal on ground (c) in paras 17 to 21 G
of the Secretary of State's decision letter. He said:

> 17. You point out that notices 1/2 show different sites from notices 3/4. They cannot, in your view, both be the right site. They cannot both identify the correct planning unit. The correct planning unit is the smaller site where the motocross activity takes place (Notices 1/2). It is the smaller site that is occupied by Ramsey Sports Ltd. All other land at Hill Farm is within the occupation of the farm partnership of R W and J P Ramsey from whom H
> Ramsey Sports Ltd hire the motocross track. The land at Hill Farm outside the motopark is in agricultural use and for no other purpose whatsoever. No breach of planning control has taken place on any of that land and to that extent Notice 3 relates to a large area of land on which the alleged breach of control has not occurred as a matter of fact. Thus, you conclude that the appeals relating to Notice 3 should succeed on ground (c).
>
> 18. The Council's response to this argument is that any part of Hill Farm

> could, in practice, be used as a motocross park. Other parts of the farm might not be so good for many reasons but if enforcement action only resulted in the closure of the existing tracks they could be moved to elsewhere on the farm causing as many problems as now if not more. If the notices can only be directed at the track itself the breach of control could go on for years as the council served one notice after another. It would make a mockery of enforcement action.
>
> 19. In my view there is no error in the identification of the large area shown on the plan accompanying Notice 3. The plan identifies a site of 112 ha. The alleged breach is described as a mixed use for agriculture and vehicular sports and leisure. Those uses do take place within the 112 ha site. The notice does therefore correctly state the breach of planning control.
>
> 20. In relation to the identification of the correct planning unit it is, in my view, important to remember that there are two different appellants, two different areas of land and two different descriptions of breaches of control. In these circumstances two planning units are involved. The smaller, 7.6 ha, site is the planning unit as far as Ramsey Sports Ltd are concerned. The company occupies that site and the breach is one of a complete change from agriculture to vehicular sports. The larger, 112 ha, site is the planning unit as far as R W Ramsey is concerned. He occupies the whole of that site. (He farms the agricultural part and he is a Director of Ramsey Sports Ltd.) The breach involved with this planning unit is that of change of part of the site to vehicular sports. I see nothing illogical therefore in the serving of Notice 3 as well as Notice 1. Such a course of action is, in my opinion, a sensible precaution bearing in mind the possible problems described by the Council (as described above).
>
> 21. I conclude that as both agricultural and vehicular sports and leisure activities have occurred as a matter of fact on the land identified by the plan attached to Notice 3 the appeal fails on ground (c).

The inspector's overall conclusion in relation to the appeals on ground (a), which were made against both notices, is contained in para 46 of the decision letter, which I need not read. In the outcome, as appears from para 51, he upheld both notices.

In his submissions Mr Bedford referred to five reported cases, dealing with the need to undertake the exercise of identifying a planning unit for the purpose of deciding whether there has been a material change of use within section 55(1) of the 1990 Act. He started very properly with the case of *G Percy Trentham Ltd* v *Gloucestershire County Council* [1966] 1 WLR 506. The case is well known and it is not necessary to refer to it in detail.

Mr Bedford next referred me to the equally well-known case of *Burdle* v *Secretary of State for the Environment* [1972] 1 WLR 1207, a decision of Bridge J (as he then was). In that case the headnote suggests, and Mr Bedford I think invited me to draw the conclusion, that the exercise of defining the planning unit could only result in one answer in any particular case. The test applied was whether there were two or more physically separated and distinct areas which were used for substantially different and unrelated purposes. Mr Bedford invited me to say that that was a test that could be applied only once in relation to any particular activity on any particular area or site. The facts there were that the land in question had a number of buildings on it, including a lean-to annexe attached to a house. The land as a whole was used as a scrap-metal and car-breaker's yard and salvage vehicle parts were sold from the site.

The local authority took the view that there had been a change in the character and degree of retail sales from the site which constituted a material change of use of the whole site. They served an enforcement notice applying to the whole site. The occupier's case was that as a matter of fact and degree, looking at the site as a whole, the intensification had not been sufficient to amount to a material change of use.

At the foot of p 1210 in the *Burdle* case Bridge J set out the inspector's conclusions as follows: A

> The legal implications of the above facts are matters for the consideration of the Secretary of State and his legal advisers but it appears to me, from the almost complete absence of reference to wholesale deliveries, that the original business was based on the scrap yard, grew out of the then proprietor's specialisation in the Austin 7, an obsolete vehicle, and would not have survived as a mainly retail business. In contrast, while sales of salvaged B spares survived, the combination of advertising with improved facilities for display, and the emphasis on new items in that display, all now support the [occupier's] claim that the annexe is a shop. But in becoming a shop a material change has taken place, without planning permission and later than January 1, 1964. Whether or not [the notice] . . . is properly directed to the whole property or to the annexe, the appeal should therefore fail on ground (d).

C

Bridge J went on to say:

> I read that conclusion as indicating . . . that the inspector was aware, although it does not appear from the report that it was raised by the parties, that there was an issue for consideration as to what was the appropriate planning unit to be considered, either the whole site on the one hand, or on the other hand the lean-to annexe, but he took the view that whichever unit one considered, there had been a material change of use, and accordingly he D thought the enforcement notice could be upheld on that footing. Speaking for myself, if the Secretary of State had adopted and endorsed that view, I do not see that such a conclusion could have been faulted in this court as being erroneous in point of law.

If one pauses there, one sees that it was Bridge J's view, concurring with the inspector, that there was a possibility that either the unit would have been E correct, that is to say the whole of the premises or the lean-to on its own. But there was, of course, only one notice. Bridge J then pointed out that the Secretary of State did not accept that view and, as appears from the report, he took the view that the enforcement notice as worded could relate only to the lean-to or the annexe. He therefore took steps to amend the notice to make that clear. Having done that, the Secretary of State went on to ask the question whether there had been a material change of use of the annexe and found that F there had been. Bridge J said in relation to that:

> Given that the lean-to annexe was the appropriate planning unit for consideration, the decision of the Secretary of State that there had been a material change of use of it was, as I think, clearly right . . .

He said that he could not accept that the Secretary of State in any way exceeded his jurisdiction in ordering that the scope of the enforcement notice G be cut down, if it was originally intended to apply to the whole site, so as to limit the ambit of its operation to the lean-to annexe. As such, that was a variation in favour of the occupiers.

So far, that is a course that did not have to be considered by the inspector in this case because he had two notices before him, one of which was applicable to the whole of the farm and the other only to the motopark. H

Bridge J then went on to consider the matter more generally. He said at the bottom of p 1211:

> But the real complaint and grievance of these occupiers is that the Secretary of State has for insufficient or incorrect reasons directed his mind to the wrong planning unit and thereby deprived them of a consideration and decision by the Secretary of State, as opposed to the inspector, of the real question which the occupiers say should have been considered, namely, has

> the change of activities on the whole site effected a change of use of the whole site which is the appropriate planning unit to be considered?
>
> For my part I am unable to accept that the reasons as expressed by the Secretary of State in his decision letter were good reasons for concluding that the lean-to annexe was the appropriate planning unit for consideration. I accept at once that whether one uses the definition of "shop" in the . . . (Use Classes) Order 1963 or the ordinary dictionary meaning of "shop", it is really an absurdity to describe the whole of this site as a shop, but what I cannot accept is that the accident of language which the local planning authority choose to use in framing their enforcement notice can determine conclusively what is the appropriate planning unit to which attention should be directed.
>
> What, then, are the appropriate criteria to determine the planning unit which should be considered in deciding whether there has been a material change of use? Without presuming to propound exhaustive tests apt to cover every situation, it may be helpful to sketch out some broad categories . . .

Then Bridge J set out his well-known three categories. He went on at p 1213 B:

> It may well be that if the Secretary of State had applied those criteria to the question, what was the proper planning unit which fell for consideration in the instant case, he would have concluded on the material before him that the use of the lean-to annexe for purposes appropriate to a shop had become so predominant and the connection between that use and the scrap yard business carried on from the open parts of the curtilage had become so tenuous that the lean-to annexe ought to be regarded as a separate planning unit.
>
> But for myself I do not think it is possible on the factual and evidential material which is before this court for us to say that that was by any means an inevitable conclusion at which the Secretary of State was bound to arrive, and that being so I do not think it would be appropriate for us to usurp his function of deciding the question, what is the appropriate planning unit here to be considered as a matter of fact and degree?

Accordingly, the appeal was allowed and the case was sent back to the Secretary of State with a direction to reconsider his decision in the light of the judgment. Willis J and Lord Widgery CJ agreed.

In this case there were two enforcement notices and each enforcement notice had to be considered on the material relevant to the challenge that was made to it. In relation to enforcement notice 1 there was no challenge other than under grounds (a) and (h), so that the occupiers were content to accept enforcement against the smaller site. They chose to challenge the propriety of enforcement notice 3, which applied to the larger site. One can see from the inspector's comments why it was in their interest to do so and why the local planning authority were concerned that they should not.

If one looks at each notice in isolation, it seems to me that, as Bridge J indicated in *Burdle's* case, one cannot fault a conclusion that in relation to its own facts the definition of the planning unit and the definition of the alleged breach of control within each enforcement notice are perfectly proper and indeed conventional. It seems to me that if one looks at each in isolation there is no fault to be found in the decision letter at all, and indeed Mr Bedford's withdrawal of the first ground of his notice of motion implicitly accepts that point.

I will not refer for the moment to the other cases, which were put before me. I go on to consider the substance of Mr Bedford's submissions. He very helpfully made them by reference to his note. The first point he took was that there was only one activity, that is the motopark, referred to in enforcement notice 1 and the same activity was referred to in enforcement notice 3. There is no question but that that contention is right, and appears from the facts found by the inspector in the decision letter in passages to which I do not need to refer.

Second, Mr Bedford said that whether that activity was in fact development depended on whether there had been a material change in the use of the land. For that purpose one had to identify the relevant planning unit. That proposition is plainly right. It is an exercise that had to be undertaken at some stage, and it was done at different stages effectively for each of these enforcement notices. I say at different stages because, the first notice not having been challenged, there was an acceptance of the local planning authority's decision to frame enforcement notice 1 in the way they did, and the decision in relation to the second enforcement notice, notice 3, was initially made, of course, by the local planning authority but confirmed by the inspector in the way in which I have indicated. There is no difficulty about Mr Bedford's second proposition.

He then continued to the third point, which was that the inspector had held that there were two planning units, one for each site as defined in each notice respectively. Mr Bedford pointed out that he justified his decision in relation to enforcement notice 3 by referring to the fact that there were two appellants, two areas of land and two different descriptions of breaches of control.

It seems to me that that is not strictly so. The inspector certainly pointed out that there were those facts. His reason for concluding that the larger site (the whole farm) was a proper planning unit for the purposes of enforcement notice 3 and the decision as to whether motopark use on part of the farm was indeed a material change of use seems to me quite clearly to be encapsulated in what he says about Mr Ramsey. He says the larger 112 ha site is the planning unit. As far as Mr Reginald Ramsey is concerned, he occupies the whole of that site. He farms the agricultural part and he is a director of Ramsey Sports Ltd. It seems to me that the inspector was properly directing that remark to a consideration of whether Mr Ramsey was in control, not only of the farm as a whole, as it were, excluding the 7.6 ha, but also, by virtue of his directorship of the company which he had formed with his brother, in practice in control of the 7.6 ha as well. Since the exercise of identifying a unit of occupation is implicity concerned with the area which is under control rather than with considerations of the nature of the tenure, in my view the inspector's approach was correct and he was entitled to hold that for the purpose of enforcement notice 3 the whole farm, including the 7.6 ha, was the planning unit.

Mr Bedford went on to make the fourth point that, as there was only one activity in effect being enforced against, the number of appellants made no difference to the identification of the correct planning unit. It seems to me that that submission is right. In relation to the smaller unit there was never any dispute, and in relation to the larger unit the matter seems to me to have been properly concluded by the inspector's decision.

Mr Bedford's fifth point was that the inspector should then have decided whether the activity was being carried on only with reference to the 7.6 ha site, in which case that was the proper planning unit, or whether it was being carried on on the whole 112 ha site, in which case that would be the proper unit. That of course is what I have to decide, so that that proposition really poses the question. In my judgment, it was lawful for the inspector to decide that the notice in relation to the large site was a valid notice and, since the validity of the notice in relation to the smaller site was not challenged, he was correct in upholding both. But the inspector does not deal expressly with the contention that it was not possible for both notices to be upheld because it was not possible on the facts to have two planning units because they overlap or the larger includes the smaller.

Mr Bedford's sixth point is that since there was only one activity the different descriptions given to the breach of planning control in the respective

A notices could make no difference to the identification of the correct planning unit. It seems to me that the descriptions of the breach, it is right to say, do not make any difference to the identification of a planning unit. The breach in relation to the small unit was that the use had been changed from agriculture to leisure. In relation to the larger unit the change was from agriculture to a mix of agriculture and leisure.

Mr Bedford's seventh proposition was that, in relation to the activity being B carried on, the inspector should either have found that the smaller site was the correct unit and judged the materiality and change against that or have found that the larger site was the correct unit and judged the materiality and change against that.

It seems to me that that is not right because the question of materiality and change in relation to the smaller unit was not raised before the inspector. The C appeal was on ground (a). I find that that proposition does not advance the argument.

I am left with Mr Bedford's eighth and ninth points, which are, in effect, if I can run them together, that at the end of the day it was necessary to decide that there was a correct planning unit, that you could not have two correct planning units. That is something that the inspector did decide by upholding both notices, but he did not give any reasons except in so far as the reasons can D be implied from the decision, as they can quite clearly. This is not a case in which the inspector's reasoning is challenged. Can one have two enforcement notices applying to the same area in relation to the same activity? I reserved my judgment to consider the implications, which were not extensively canvassed in argument on either side.

Mr Kent, for the Secretary of State, identified the issue whether it was only possible to have a single planning unit. He pointed out, rightly, that there is no E clear authority on the subject. The question of what is a planning unit and the underlying reasoning for the planning unit exercise is entirely law made by the courts.

Without doing him an injustice I can shorten my account of what he had to say in this way. He says in effect that although the development of the planning unit as a tool has been considered, generally speaking, in relation to the validity of splitting up the site and serving a number of enforcement F notices as distinct from serving a single enforcement notice in relation to the whole site, it does not follow from either the case of *Burdle* or the later case of *de Mulder* v *Secretary of State for the Environment* [1974] QB 792 that it is necessary to limit enforcement procedure to one notice for each site. It is permissible to serve more than one enforcement notice relating to the same site and to the same activity on that site when the site on which the contravening G activity is carried on can, in fact, be regarded as being comprised in more than one planning unit.

It seems to me that there is no justification for remitting the case to the Secretary of State. My reasons are these. First, that an enforcement notice has to be considered initially in the light of the provisions of sections 172 and 173 of the 1990 Act. Having looked at those provisions, I can see no requirement in those provisions to limit the enforcement procedure to a site or sites that do not H overlap one another.

Second, considering the matter at the appeal stage, I have looked at sections 174 and 175; and there I can also find no indication of a requirement of exclusivity. There is no indication that several enforcement notices cannot properly be served in relation to the same site or area of land, or indeed in relation to one activity on a site that is defined independently of any other site or that is defined in such a way as to form part of or be overlapped by or overlap a second site.

A Third, it seems to me that there are good practical reasons not to impose any strict requirements which are not expressly derived from the statute. This case is a good example of seeking in practice to be able to control by enforcement procedure the use not simply of the small site but also of the large, because there is a distinction in the occupation albeit the inspector has found that Mr Ramsey is effectively in control of the company. Using the inspector's language, the company hired the motocross track. Mr Bedford told me that B that was under a licence. There is ground for saying that the company are in control of the smaller site — that was never challenged — and there is equally ground for saying, as the inspector did, that Mr Ramsey was, in practice, in control of the whole farm, including that site. Quite plainly the apprehensions of the local planning authority as to the possibility of shifting the motor-sport activity around within the boundaries of the farm were apparent to the parties at an inquiry and to the inspector. I regard that as good practical justification C for upholding both notices, as he did, if it is lawful to do so.

Is there any constraint imposed by the decided cases? It seems to me that there is no implication in those decisions which would require me to hold that it is lawful and necessary to serve only one notice. There are constraints which appear from those decisions; for example, in the *de Mulder* case it was pointed out that the choice exercised by the local planning authority of whether to split up a single holding into a number of smaller units for the purpose of serving D enforcement notices is open to the authority. But they may run into the difficulty that appeared in the *de Mulder* case, which is plainly something that the courts have to guard against, namely that the effect of splitting up the site will restrict a landowner's rights in a way that is not justified.

The enforcement notice procedure should not be used inadvertently to the effect that a landowner's rights are cut down, because it is designed only to be utilised against development which is in breach of control. E

I have not been able to find any other constraint indication in the *de Mulder* case, or in *Burdle*, or in the other two cases to which I was referred, namely *Johnston* v *Secretary of State for the Environment* (1974) 28 P&CR 424 and *Rawlins* v *Secretary of State for the Environment* [1989] JPL 439*.

In those circumstances, I next consider the effect of upholding both notices in this case. If the smaller site is considered, it seems quite clear that the effect of the notice that the motopark use must cease will be to preclude any future F use of the 7.6 ha for the motor-sport and leisure activity. If one looks then to the effect of the notice in relation to the farm as a whole, if that notice is upheld there will be no lawful motopark use anywhere within the farm. But that does not restrict the rights of the landowner because, except within the limits of, for example, the general development order, his use of that land is an agricultural use and if other activities are introduced which are sufficiently substantial to G constitute a material change of use then he would plainly be carrying out a development in breach of control. Enforcement notice 3 has no improper restrictive effect in relation to the large site. There is a possible difference in the effect of the two notices that it may be that some relatively trivial motopark activities could be carried on on the farm without constituting a breach of control whereas, it seems to me, no motopark activities could be carried on on the small site as such. But that difference is not a constraint upon the owner's H rights in any way. By owner in that context I mean the rights of the person in control of the land, whether it be Mr Ramsey or the company.

The final consideration that seems to me to be relevant, although it was not referred to, is that, of course, if both notices are upheld, it would be possible for a prosecution to be based in relation to the contravention of both notices if

*Editor's note: This is a reference to the report of the judgment of Sir Graham Eyre QC. His decision was upheld on appeal: [1990] 1 PLR 110.

A the contravening activity took place upon the 7.6 ha site. That would be a clear offence in relation to each notice. If any planning authority were to seek to prosecute under both notices in respect of the same alleged contravening activity, it seems to me that that would be unjustified and that it would not be accepted by any court. Although there is that theoretical possibility, that is not a true example of double jeopardy because the notices themselves do not in any way put a landowner in jeopardy. It seems to me that the possibility cannot
B provide a material consideration for saying that in these circumstances either notice must be unlawful.

In those circumstances, and for those reasons, the inspector's approach and what I regard as his reason for upholding enforcement notice number 3 were, in my judgment, correct. In relation to the other notice, its validity was not called in question before the inspector except in so far as there was an appeal
C under ground (h). There is no challenge to that part of his decision.

In those circumstances I can find nothing wrong with this decision and the appeal will be dismissed.

Appeal dismissed with costs.

D

NOTER-UP

Safeway Properties Ltd v Secretary of State for the Environment
[1990] 3 PLR 87

E The appeal against the decision of Otton J was allowed.

Coram:

Court of Appeal
Parker and Taylor LJJ and Sir George Waller
June 14 1991

For Product Safety Concerns and Information please contact our EU representative GPSR@taylorandfrancis.com
Taylor & Francis Verlag GmbH, Kaufingerstraße 24, 80331 München, Germany

www.ingramcontent.com/pod-product-compliance
Lightning Source LLC
LaVergne TN
LVHW021726021026
849161LV00004BA/735
* 9 7 8 0 7 2 8 2 0 1 6 7 5 *